W9-BNW-618

*f***P**

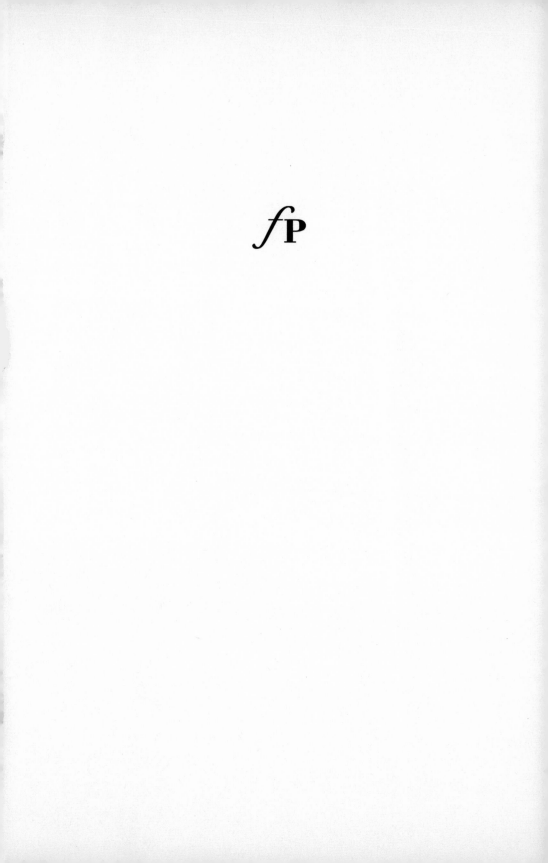

BORN BELIEVERS

The Science of Children's Religious Belief

JUSTIN L. BARRETT, PH.D.

Free Press
New York London Toronto Sydney New Delhi

*f*P
Free Press
A Division of Simon & Schuster, Inc.
1230 Avenue of the Americas
New York, NY 10020

Copyright © 2012 by Justin Barrett, Ph.D.

All rights reserved, including the right to reproduce this book or portions thereof in any form whatsoever. For information address Free Press Subsidiary Rights Department, 1230 Avenue of the Americas, New York, NY 10020

First Free Press hardcover edition March 2012

FREE PRESS and colophon are trademarks of Simon & Schuster, Inc.

For information about special discounts for bulk purchases, please contact Simon & Schuster Special Sales at 1-866-506-1949 or business@simonandschuster.com.

The Simon & Schuster Speakers Bureau can bring authors to your live event. For more information or to book an event contact the Simon & Schuster Speakers Bureau at 1-866-248-3049 or visit our website at www.simonspeakers.com.

Designed by Ellen R. Sasahara

Manufactured in the United States of America

10 9 8 7 6 5 4 3 2 1

Library of Congress Cataloging-in-Publication Data

Barrett, Justin L.
Born believers: the science of children's religious belief / Justin L. Barrett
p. cm.
Includes bibliographical references and index.
1. Faith—Psychology. 2. God. 3. Child psychology. 4. Children—Religious life. 5. Psychology, Religious. I. Title.
BF723.G63B37 2012
200.1'9—dc23 2011039581

ISBN 978-1-4391-9654-0
ISBN 978-1-4391-9657-1 (ebook)

To Skylar and Sierra

Acknowledgments

THIS BOOK HAS BEEN several years in the making with the assistance of many friends and colleagues to whom I am sincerely grateful. I thank Rita Astuti, Lisa Barrett, Pam Barrett, Jesse Bering, Pascal Boyer, Kelly Clark, Emma Cohen, Miguel Farias, Paul Harris, Laura Hewitt, Carl Johnson, Deb Kelemen, Bob McCauley, Michael Murray, Josef Perner, Jeff Schloss, Richard Sosis, Henry Wellman, and Harvey Whitehouse for feedback, suggestions, discussions, and anecdotes that have proven valuable. I am indebted, too, to my collaborators on various experiments described here, particularly Emma Burdett, Amanda Driesenga, Frank Keil, Nicola Knight, Amanda Johnson, Roxanne Newman, and Rebekah Richert. My agent, Esmond Harmsworth, provided sage advice at many turns and helped considerably in improving the manuscript. Sherry Barrett, Emma Burdett, and Tenelle Porter soldiered through earlier, rougher versions of the book and helped me avoid terrible pitfalls. (If I have managed to step in some, the blame is mine and not theirs.) Donna Loffredo and the Free Press/Simon & Schuster staff were very supportive and constructive. Finally, I thank Jean-Luc Jucker for his invaluable assistance in manuscript preparation. I fear I have overlooked other critical players here and beg their forgiveness, assuring them that I am grateful for their contributions as well. The John Templeton Foundation supported this research. Thank you for taking the time to read.

CONTENTS

BORN BELIEVERS

Introduction

On the Train to Jaipur

THE HOT SEASON HAD begun, and the sun bleached the barren landscape outside the train from Agra to Jaipur, India. Inside, amber dust eddies scampered down the aisles and among the rows. I sat uncomfortably on the squeaky, sticky, turquoise-colored vinyl seat and glanced at my fellow travelers. Nearby was a middle-aged man dressed in a single bright orange cloth draped over one shoulder like a toga. In contrast with his bald crown, grizzled hairs carpeted his exposed shoulder, arms, and legs.

"He's a saint," a well-dressed man across the aisle commented, noticing my gaze. The clean-looking Brahmin with a thick black mustache initiated a lengthy conversation with me, offering explanations of various aspects of Hinduism. Eventually our talk turned to my purpose in visiting India. I went to India as a psychologist studying people's concepts of gods.

"What have you discovered?" he asked. Being a young scholar convinced of the importance of not drawing conclusions without good evidence, I was reluctant to claim that I had discovered anything—at least not yet—but his inquiry required an answer. I told him that my first set of experiments on God concepts seemed to show that adults had a hard time using their stated beliefs about God in certain contexts. For instance, though denying that God has a particular location, most of the participants in my studies appeared to understand stories about God by assuming that God is in one place at a time, much like a human (more on these studies in Chapter 6). But I had also been working on new experiments with young children that revealed that they had much easier times thinking about God than I had anticipated. Adults surprised me with the difficulty they had using their God concepts, but children used God concepts with ease.

I assumed that the Brahmin would nod with that I-have-no-idea-what-you-are-talking-about-or-why-anyone-would-care-but-I'm-too-polite-to-say-so smile that several friends and family members had indulgently offered me previously. Instead, the man smiled knowingly and asked confidently, "Do you want to know why?" Sure. He explained to me that on death, we go to be with God and later are reincarnated. As children had been with God more recently, they could understand God better than adults can. They had not yet forgotten or grown confused and distracted by the world. In a real sense, he explained, children came into the world knowing God more purely and accurately than adults do.

Since that train ride, I have conducted numerous additional studies on religious beliefs, and colleagues in my field, the cognitive science of religion, have discovered more evidence that children have a natural affinity for thinking about and believing in gods. Perhaps surprisingly, the evidence to date suggests that as the Brahmin indicated, children show remarkable natural affinities for thinking about and believing in gods. This book shows how children naturally develop minds that encourage them to embrace belief in the god or gods of their culture. People may practically be *born believers.*

A comparison may be helpful. Maybe you've heard a pronouncement that someone was a "born singer" or "born artist." My mother recalled that on the birth of my brother, the doctor declared him a "born basketball player." (He wasn't.) Children aren't born singing or painting or shooting a turnaround jump shot, but these expressions mean that babies are born with capabilities that will—if given minimal opportunities and cultural support—unfold in such a way as to produce mastery in singing, art, or basketball. In a related (but not identical) way, essentially all human babies are born talkers—destined to acquire language—and born walkers—naturally going to learn to walk. In a similar way, children are born believers in some kind of god.

Children are prone to believe in supernatural beings such as spirits, ghosts, angels, devils, and gods during the first four years of life due to ordinary cognitive development in ordinary human environments. Indeed, evidence exists that children might find especially natural the idea of a nonhuman creator of the natural world, possessing superpower, superknowledge, and superperception, and being immortal and morally good. I call this type of supercreator god *God* for short. That's right: children's minds are naturally tuned up to believe in gods generally, and perhaps God in particular.

At this suggestion about the naturalness of religious belief, perhaps you are already considering an alternative account of childhood religiosity. Maybe you have seen video footage of Islamic schoolchildren in traditional garb, ritualistically repeating sections of the Quran over and over for hours every day, through what appears to outsiders as a system of coercive programming. Maybe, too, you have heard of monastic traditions that virtually imprison youths for years of their lives, disallowing them contact with the outside world until they have fully conformed to the values and behaviors of their elders. These examples convince some observers of religious development that what is needed to convince children of doctrinal beliefs is thorough, unmerciful, systematic brainwashing.

A more measured version of this common explanation for why children seem to so readily believe in gods might be called the indoctrination hypothesis. In short, children believe because their parents and other important adults in their community teach them to believe: they indoctrinate them. And as children do not really have the mental resources to think for themselves, they blindly go along with what these adults say. After all, to disagree could be dangerous.

Some very smart people mistakenly think indoctrination is the whole story. At a conference where Pascal Boyer, another cognitive scientist of religion, and I presented what might be called the "naturalness of religion" thesis—the idea that the natural architecture of human minds in ordinary environments makes belief in gods entirely expected—we were asked a question to this effect: "Isn't it the case that you can teach kids to believe in any crazy thing as long as it can't be disproven and you punish them if they don't believe?" Boyer put his answer this way. If you told a child that Dick Cheney was made of green cheese except whenever anyone looked at him, it would not matter how much you beat the kid if he did not believe or how much you threatened him with eternal damnation. The best you would get is for the child to pretend to believe that Dick Cheney is made of green cheese, but you could not coerce or indoctrinate the child into this belief, even though it could never be disproven. No doubt Boyer had in mind research by developmental psychologists such as Henry Wellman and Paul Harris that shows even preschoolers understand the difference between reality and fantasy.[1] Preschoolers know that an imagined pony or monster cannot be seen or touched by someone else, even if it provokes strong feelings of comfort or fear. In this way, preschoolers are not so different from adults who are getting emotionally involved while watching a movie: we know it is not real, but it can still get our heart pounding or move us to tears.[2]

Religious ideas are very different from pretend or fantasy. Since I have begun giving public lectures on the science behind this book, many people have relayed to me accounts of how readily their children have embraced religion or how difficult it has been to dissuade them of belief in God. A former coworker of mine told me about her three daughters, the oldest only eight years old. "I'm a Christian but my husband is an atheist, so we agreed not to push our kids in either direction," she explained. "But it doesn't seem to matter. All three girls believe in God, and not just a bit. Sophie, my oldest, has rows with her dad and tells him he's wrong about God not existing."* An atheist mother from Oxford, England, was amazed to discover that her five-year-old son had a firm belief in God against her best efforts. Unbelieving parents in Indiana reluctantly let their kindergartner go to vacation Bible school, and she came home expressing a desperate desire to continue learning about God. A Danish colleague (to whom I return later) discovered his little girl had casually contracted a strong case of theism even in one of the most secular societies on earth. These and many other anecdotes are not my reasons for saying that children are born believers, but they suggest something beyond happenstance or indoctrination is going on here. Why is it so much easier to get kids to believe in some kind of god than other beliefs such as believing in the virtues of broccoli, that their great-aunt isn't really scary, or that there isn't any god?

Parents of young children (and especially teenagers!) know that they cannot just program their kids' beliefs. Sometimes we can "indoctrinate," but often it does not work. For instance, comedian Julia Sweeney has tried to raise her daughter as an atheist, but apparently it hasn't been easy. In this amusing extract from a *San Francisco Chronicle* interview, Sweeney explains:

> I said God is this idea of a big man who lives up in the clouds and he created everything. And she [Sweeney's daughter] goes, "Well I believe that!" And I go: "Well yeah, because it sounds like a cartoon character. But the truth isn't that, and I'll tell you the truth."

---

* Sophie is a pseudonym, as are all names of children used in the book.

> And then I actually teach her about evolution, and she asks me about it all the time as a bedtime story. She'll say, "Tell me about how people weren't here when the dinosaurs were here." And then we'll go over it again. I don't know how much of it she really gets, but she likes the story. And then, she's kind of over it now, but she would go, "I believe in God at school, but when I come home I don't."[3]

This excerpt illustrates how difficult it can be to indoctrinate children away from religious belief. Perhaps the daughter negotiates her mother's opposition to God by conceding that she just will not believe in God at home. Sweeney's persistent attempts to indoctrinate her daughter against belief in God face serious challenges because of children's natural tendencies toward religious belief. Children are not ready to believe all ideas equally.

The indoctrination hypothesis persists because often people underestimate how much information children are born already having or are predisposed to acquire easily and rapidly. We often carry around the assumption that babies' minds are like empty containers waiting to be filled, and it does not matter what you put in their heads as long as it is not too much. Under this view of human minds, learning to believe in gods or learning to believe in subatomic particles is pretty much the same process. The only differences might be opportunities for learning and motivation. Children have more people around telling them about God than telling them about subatomic particles, and maybe children have more motivation to learn about God because the idea of God gives them comfort on dark, stormy nights.

I will return to this issue, but here let me dispel the notion that human minds are like empty containers simply waiting to be filled.[4] This view ignores that human minds have a considerable number of natural tendencies that allow them to solve problems important for their survival and life concerns. From birth, human minds acquire and handle some kinds of information more efficiently than others. For instance, research indicates that within hours of birth, full-term babies can already imitate some facial expressions, such as pursed lips or a gaping mouth. If babies see someone stick out a tongue at them, they are more likely to stick out their tongue.[5] Such actions require that babies somehow recognize faces and what those faces are doing, and they map that action onto their own facial muscles—even though they have never seen their own face! From birth, then, humans are excellent at recognizing human faces. By adulthood we can identify, remember, and distinguish among thousands of faces effortlessly—a feat that sophisticated computers find unwieldy. In contrast, solving multiplication problems with three-digit numbers requires considerable effort and targeted education. Computers from decades ago could easily solve such problems.

What the contrast between face recognition and multiplication shows is that human minds are specialized to handle some types of information and problems more readily than others. Likewise, not all ideas or beliefs may be acquired just as easily as any others. We find many of the ideas that physicists study more difficult to learn than many religious beliefs because they really are more difficult for our minds. Our minds find them more foreign, further away from what our minds naturally do, than many religious ideas.

Regardless of culture and without need for coercive indoctrination, children develop with a propensity to seek meaning and understanding of their environments. Given the way their minds naturally develop, this search leads to beliefs in a purposeful and designed world, an intelligent designer behind the design, an assumption that the intentional designer is superpowerful, superknowing, superperceiving, and immortal. This designer does not need to be visible or embodied, as humans are. Children readily connect this designer with moral goodness and as an enforcer of morality. These observations in part account for why beliefs in gods of this general character are widespread cross-culturally and historically.

Think of it this way. Perhaps you remember a preschool shape-sorter toy that is a nearly round, hollow, red and blue plastic object with lots of different yellow shapes that fit into matching holes. Ordinary child development provides children with a number of conceptual holes that have particular shapes. One of these holes is a god-shaped hole. Children are naturally ready to receive the shape—the cultural idea—that fits well into the hole: gods of various sorts. Some gods fill the hole better than others, but many fit just fine. In playing with the toy, however, you might remember that you can cram the wrong shape into some of the holes because it is a close enough fit. Similarly, the god-shaped conceptual hole can be filled with beings and ideas other than gods such as human idols, governments, or a personified Natural Selection or Chance. To put these misfits into the holes takes a little shoving—extra conceptual work—but they can be forced.

In the following eleven chapters, I tell the story of how children develop their beliefs in gods and highlight some of the scientific evidence that supports it. I cannot provide all of the evidence but offer enough to make my case compelling. The scientific research is new and ongoing, but it points to a clear general story of childhood belief.[6]

In Chapter 1, I describe how, from the first year of life, children show signs that they treat intentional beings—agents—in importantly different ways from inanimate objects and pay considerable attention to them. Without this propensity, children believing in gods would be as widespread as children believing in imaginary numbers—extremely rare. I then present evidence in Chapters 2 and 3 that children come into the world with a tendency to see order, purpose, and even intentional design behind the natural world, as if everything in the world had a particular function and had been intentionally ordered by someone for that purpose. Given their attention to agents and this tendency to see intentional design in the natural world, positing an intentional creator—a god—is not far behind.

In Chapters 4 and 5, I describe experiments conducted not only with American and British children, but also Greek, Israeli, Maya, and Spanish children that suggest children have a natural leg-up on predicting what God knows, sees, hears, and smells before accurately predicting the same for humans. Children also begin with an assumption that others will live forever and have to learn it is not true of humans and animals.

In Chapter 6, I begin discussing the implications of the science by summarizing what the science of childhood religion reveals to be *natural religion*—the sort of religious beliefs children seem to gravitate toward naturally. Given my argument that children are prone to believe in gods because of the way their minds develop in the first several years of life, one might ask if I think that belief in God is childish or infantile. Sigmund Freud answered this question in the affirmative. I share my own answer to this question in Chapter 7. I address the indoctrination hypothesis head-on in Chapter 8 using the research presented in the previous chapters and additional considerations. If children are born believers and religion is so natural, how do we account for atheists? Is atheism unnatural? These topics are addressed in Chapter 9. Chapter 10 provides a discussion of whether parents and other teachers and caregivers should teach children about God and religion. Is it good for children, or a form of abuse, as several new atheists have suggested recently? In Chapter 11, I offer some suggestions for how to healthily and effectively encourage children in their development, no matter your religion or lack thereof.

I hope that this book will whet your appetite for the burgeoning new psychological and evolutionary study of religion. At the end of the book are the source notes, a list of further readings, and the index.

PART ONE

The Evidence

One

Secret Agents Everywhere

If you ever visit Oxford, and I hope you will, I recommend that once you finish walking beside the dreaming spires and ancient college quads you visit the Pitt-Rivers Museum, the site of the university's anthropological collection. A visit to the Pitt-Rivers will make you feel as if you are rummaging through humanity's attic.

A popular exhibit with the nonsqueamish is a glass case labeled "Treatment of Dead Enemies." Inside are mummified and shrunken heads, some with sharpened sticks and crude blades stuck through them. A neighboring case houses scores of beautiful and grotesque figurines, many part human–part animal, some with multiple limbs. As you continue wandering, you will find additional strange items from all over the world:

bug-eyed voodoo dolls, hand-carved amulets, and tiny coffins holding mummified cats. What ties many disparate items together, including Oxford's famous dreaming spires, is religion: each of these unique, occasionally bizarre exhibits has a religious significance in the culture they come from. They are visual reminders of the diverse and nearly ubiquitous presence of belief in gods across times and cultures.

The vast majority of cultures, as well as the vast majority of people, believes in some sort of god or gods. If we count as gods all willful beings with some special property that humans or animals do not have (such as being invisible, immortal, or made of bronze), belief in gods has occurred in every age and in every culture. So many cultures are religious that religion of some sort seems to be a natural human expression.[1]

Similarly, almost everyone, no matter their culture or the beliefs of their parents, goes through a period of affirming the existence of one or more supernatural beings. Perhaps you have a childhood memory like this:

> The scene is a children's sleepover, and the parents have turned off the lights and gone to bed. Tired but excited kids try to frighten each other with whispered macabre stories until one child knowingly asks, "Do you want to see a *real* ghost?" Everyone falls silent for a moment, and then one cocksure kid says, "Yeah. How?" "All you have to do is go into the bathroom alone with the lights out, close the door, and say 'Bloody Mary' three times. Then she'll come to you in the mirror. But you'd better get out fast or she'll kill you!"

Many children have tried it and many have seen her—or at least think they have seen her. Others refuse to even try. Some of the special beings children believe in are imaginary friends, others are scary ghosts and monsters, some are benevolent Santa and elves, and many are what adults recognize as spirits or gods—the kind of beings that are part of shared religious systems.

Two Ways of Being Natural

Some ways of thinking or acting are so automatic to us, so easy, so fluent, that we can't imagine not having them. In fact, it is almost impossible for us to imagine a person growing up and *not* learning certain abilities. The word we use for this is *natural.* Philosopher Robert McCauley calls this kind of naturalness *maturational naturalness* to emphasize that it comes about as a normal part of growing up and maturing.[2] Learning to walk is maturationally natural. Understanding that you have to touch a solid object to make it move (as in picking up a coffee cup) is maturationally natural. Developmental psychology tells us that using your native language, recognizing family members' faces, and adding single-digit numbers are maturationally natural in McCauley's sense. We usually acquire these maturationally natural abilities so early in life that, as adults, we do not remember not having them. People the world over have them because these abilities do not require special training, explicit instruction, or special tools or other artifacts to acquire.

In contrast, we get some abilities through special training, instruction, using special tools, and lots and lots of practice.

Consider riding a bike. Once you've mastered riding a bicycle, your body seems to just know how to do it. You don't have to think about it—to consciously remember how to balance, steer, and control a bicycle's speed. You just do it. But you remember how difficult, scary, and frustrating it was to learn how to ride a bicycle. I remember wearing a motorcycle helmet to protect myself from repeated crashes on the gravel driveway and into trees and hedges, my dad giving me direct instructions about how to ride and helping me get started. At one point, it was not so automatic, not so natural.

Years of practice, instruction, and correction yield practiced naturalness, as McCauley terms it. Effortless driving of a car, mastery of algebra, and reading are all cases of practiced naturalness. If you do not get instruction and practice a lot, you do not acquire mastery, and lots of perfectly intelligent and capable people never do acquire mastery of these. Special cultural conditions are required for this kind of naturalness.

McCauley uses the term *naturalness* to refer to both classes of capabilities because we find them easy, automatic, and fluent. I do not have to concentrate to read a road sign. If I see one, I read it automatically. Likewise, I usually do not have to do any careful concentrating to put a sentence together when I speak to someone. Both reading and speaking seem natural in the sense of being easy and automatic. But this apparent similarity masks a hidden difference. Speaking is essentially inevitable for us as normally developing humans. Parents do not tell us how to speak and give us special speaking tools or lessons; they just speak around us, and we pick it up naturally. Reading, however, is not inevitable. Something special—a writing system, printed words, instruction, and lots of conscious, deliberate practice—must be added to ordinary natural language skills to get literacy.

I will break from McCauley's terminology and refer to the nearly inevitable capacities, thoughts, and practices as *natural* traits *or nature* and those that require special conditions, training, or practice as *expert* traits or *expertise.* We are natural language users but have to acquire expertise in reading and writing. Walking is natural, but doing ballet requires expertise.

This use of *nature* (versus *expertise*) tracks closely to common use and carries the additional benefit of allowing for some ideas, practices, or competencies to be more or less natural. Being able to add 1 + 1 might be fully natural, and adding larger sums might be mostly natural, but doing calculus is very unnatural. And many basic religious thoughts and practices are on the natural end of the continuum, while the religious thought and practice as we see in adults certainly involves a degree of expertise overlaid on a solidly natural foundation.

We sometimes make the mistake of thinking that for an ability to be natural, it must be somehow built into our biology from birth, or hardwired into our brains. Popular news stories about genes for anything from disease resistance to hair color to intelligence reinforce this misunderstanding. But just because we have a biological disposition toward a trait does not mean it will develop without the right kind of environment, and just because something is not built in does not mean it is not nearly inevitable as a part of human development. What we can more sensibly say is that given a certain kind of biological endowment *and* the ordinary sort of world we are typically born into, we will typically develop certain properties and attributes. These sorts of traits—those that are almost inevitable because of our biology plus the regular sorts of environments people grow up in—are natural traits. We can leave the "hardwired" talk to electricians.

Belief in gods of some sort or other, and maybe a supreme capital G God in particular, may be largely natural in this sense: biology plus ordinary environment, no special cultural conditions required, a predictable expression of our biology's development in a normal environment—but not be biologically determined.

Exciting new research points to just those systems of the human mind that make us born believers. For the next several chapters, I identify these early-developing systems and explain how they make belief in some kind of god almost inevitable. Children are not susceptible to religious thought because they do not yet know the way the world works. Rather, they have strong propensities to believe in gods because gods occupy a sweet spot in their natural way of thinking: gods are readily and easily accommodated by children's minds and fill some naturally occurring conceptual gaps rather nicely. There are specific early-developing mental systems that undergird childhood religious belief. And the mental system that divides the world into those things that act from those things that can only be acted on is one of the foundations of being a born believer.

Separating Who? from What?

Fantastical films like the *Harry Potter* movies or *Bedknobs and Broomsticks* fire up the imagination and give children all sorts of ideas, not the least of which is the delicious whimsy that children might be able to get furniture to fly, or even sing and dance, by casting a magic spell. Maybe we can use the Force to move objects the way a *Star Wars* Jedi does. Captivated by the events of a magic-filled movie, a child might try to get ordinary inanimate objects such as beds, books, chairs, stones, and trees to move around or change shape by talking to them, persuading them, or ordering them. But soon he will learn that no matter how much he tries, he cannot move objects through the power of mind alone.

A similar lesson is that to get people to move, it is best to ask them rather than attempting to get them to act through physical contact, pokes, prods, shoves, and punches. The difference between a chest of drawers and Uncle Billy is the difference between an inanimate object and a being that has intention and purpose. A cupboard will not move unless someone pushes it or makes it move, whereas Grandma can walk in and out of the room whenever she wants.

Fortunately, children from infancy show signs that they know the basics of how things work in the world. Ordinary objects do not launch without being touched, magically teleport from one place to another, or simply vanish from existence.[3] Children find magical tales delightful precisely because they know the world does not really work that way. We know babies already have a pretty good grasp on how bedknobs and broomsticks really behave from ingenious techniques that experimental psychologists have developed over the past thirty years for peering into the minds of babies.

To see what babies have learned about the world by a particular age, we need to perform experiments on them to see if they are surprised when things do not work out the way they are supposed to. Because babies cannot directly tell us and surprise can be hard to measure (What would the unit be—the gasp or gurgle?), scientists use changes in how long a baby looks at something as an indicator of interest or surprise. If a baby has been looking at a display and gotten bored, she will let you know by looking away, squirming, and fussing—sort of like adults do when they are bored. But, if something new or surprising is then presented, the baby gets interested again and gives the new display a good, long look. You can almost read the surprise on their faces.

One characteristic experiment was conducted by developmental psychologist Renée Baillargeon and colleagues.[4] Baillargeon's team showed two-and-a-half-month-old babies a cylinder rolling down a ramp and then coming to a halt when it struck small fixed objects dubbed *stoppers*. Nothing strange or surprising here: any object in motion will come to a stop if an obstacle blocks it.

Once babies were tired of looking at this event, or habituated (as measured by looking away from the display), the experimenters presented a slightly changed display: a toy bug on wheels at the bottom of the ramp. Sometimes this toy was placed a small distance away from the stoppers so that the rolling cylinder would not strike it, and so the toy did not get launched. Other times, experimenters placed the toy right next to the stoppers in a way such that the rolling cylinder would strike the toy. But the toy bug still did not launch into motion. From an adult perspective, this appeared to be a surprising violation of the way physics works. But would these two-and-a-half-month-olds have similar expectations? The experimenters found that the babies looked significantly longer at the display in which the toy should have been launched but was not. This finding suggests that just like adults, babies expect objects to move if another object collides with them. So two-month-olds appreciate that physical contact can launch an object. Other experiments have shown that babies also know that ordinary objects (not agents) do not launch themselves.[5]

Experiments like these give developmental psychologists confidence that five-month-olds (and perhaps even younger babies) "know" that blocks, balls, shoes, and toys have to be contacted in order to start moving, and that when they are contacted by a moving object, they tend to move. A large body of experimental evidence demonstrates that infants in the first five months of life know a lot about the core properties of common solid objects.[6] A baby who sees a shoe knows that the shoe moves together as a whole, bounded object; knows that it must move on a continuous, unobstructed path (instead of jumping from one place to another or passing through solid objects); and knows that it must be physically contacted—pushed by something else—in order to move. Later in the first year, babies show awareness that objects must be supported to keep from falling. These may sound like mundane achievements, but they are of critical importance, as they structure the physical world and make navigation and interaction with the world possible. They also help establish what happens due to ordinary causation versus supernatural causation. By *supernatural* here, I mean violating our natural expectations.

If as a baby I did not know that solid objects cannot pass through one another, I might try to walk through a closed door instead of opening it first, or I might try to reach through a cupboard door to get a toy. If I did not understand that physical objects require support, I might place my cup of juice in midair expecting it to hover. If I did not know that ordinary objects have to be contacted to be moved, I might never learn physical cause-and-effect relations such as when a rolling ball knocks over a vase or a bumped chair falls. I might also attempt to move objects without physically contacting them but by gesturing or talking to them or thinking about them, or I might have no idea how to move them and give up altogether. This collection of principles that we automatically use to think about physical objects has been termed *naive physics*—a bit like ordinary physics but a lot simpler and more natural, something we do not need to be taught.

So experiments with babies tell us that babies expect ordinary objects to behave in the ordinary ways we adults expect them to behave. Babies would find flying beds and brooms surprising. Further experiments, however, give us reason to think that babies make an exception for humans and other agents, distinguishing between the doers and the done-tos, the whos and the whats. They know that the regular rules of the world do not apply to the whos.

The difference between inanimate objects and beings called *agents* is a critical one for children to master, and the failure to do so—such as in the failure to know the difference between a boulder and a bear—can be life threatening. By *agents* I mean to include people and any other beings we understand as not merely reacting to their environment but intentionally acting on it. People or human beings are not the only agents children think about or interact with. Dogs, cats, and other animals might be considered agents. Computers, too. Ghosts and space aliens would be agents. Gods are agents, whether they are sacrifice-hungry volcanoes, weeping statues, or unseen cosmic spirits. For young children to make sense of gods, they must understand the difference between agents and brute physical objects.[7]

As children begin to grasp the critical distinction between agents and objects, they might think of ordinary, inanimate objects as the furniture of the world. These objects do not act, but only react or are acted on. A chair does not move itself around the table. An arrow in flight cannot decide to suddenly reverse course. In contrast, agents are the beings that can move the furniture and themselves.

The Importance of Understanding Agents

A couple of years ago I acquired a lop-eared rabbit named Pug. Pug was not a complete flight of fancy; he came to my office because I needed a rabbit to star in some videos that I was making as part of my work. Pug has spent lots of time visiting my office to become used to human interaction and would not behave in a typically skittish rabbit manner. He has perfected paper shredding and a few other destructive skills, and though he is comfortable around people, he has not fully mastered the trick of distinguishing between agents that have minds and nonagents. If people sit near him on the floor, Pug crawls all over them in search of food, as though they were furniture or part of the landscape. He bumps feet and tugs on them as if they were sticks. In contrast, he noses, circles, and fawns over a large pink balloon as though it were his girlfriend. These are cute behaviors in a rabbit but would be deeply disturbing in a human. Fortunately, babies are much more clever than my rabbit and readily understand the difference between agents and nonagents. Specifically, babies seem sensitive to several important features of agents that make them ready to understand humans and animals as agents, but make them receptive to gods as well:

1. Agents can move themselves and other things.
2. Agents act to attain goals (instead of just moving arbitrarily).
3. Agents need not resemble humans.
4. Agents need not be visible.

This distinction between ordinary objects and agents is so important that a glance at some representative experiments in this area of research might be helpful. The average baby performs some fascinating experiments of her own at home, keeping careful mental notes of the results (although these experiments result in Mom or Dad having to wipe things up). Here's how babies learn these properties of agents and how we know they are learning them.

Humans Can Move on Their Own: The Hat and the Bell Experiment

A friend of mine told me about his nephew, Kyle, who likes to spend time in a car seat while his uncle drives him around. When Kyle drops his juice cup on the floor, he yells and kicks the back of the driver's seat in response. Kyle knows it will not help to scream, cajole, or persuade the sippy cup. Yet it will usually work to apply such techniques to the human driver, who, when the car is not moving, usually gets out, hands the cup back to Kyle, and then resumes the journey.

To enjoy normal social interactions (and also to have religious beliefs, as I describe later), children have to understand that intentional agents such as humans do not have to be touched in order to make them move. They can move on their own. If I want you to pass the salt at the dinner table, I do not grab your arm and move it to the salt shaker and then use your arm as a lever to drag the shaker across the table. I simply say, "Please pass the salt." When is it that babies recognize that contact is unnecessary for interacting with agents? The first clues are the different ways in which babies—maybe in the first three months of life—act differently in the presence of people as compared with objects—for example, by cooing and smiling more.[8] There are also some strange but ingenious experiments that help demonstrate infants' knowledge of agents.

If I am wearing a crazy hat that has a bell on top of it, you could get me to rattle the bell either the normal way (without contact) by asking me to shake my head, or you could grab my shoulders and shake me so that my head and hat move. But if the crazy hat is sitting on a ball (instead of on my head), you have only one option. You have to physically move the ball. Asking the ball to shake the hat would be silly and pointless. Developmental psychologists have used just such a contrast to test whether six-month-old babies appreciate these distinctions. By six months, babies prefer to watch the normal causal event, such as a woman asking another woman to shake her head to rattle a bell on a hat, instead of strange ones such as a woman physically shaking the other woman's hat on her head to ring the bell.[9] Similarly, six-month-olds prefer to watch a woman physically shake the same hat when on a ball as opposed to the woman asking the ball to shake the hat and bell. Though we do not know why in this case babies prefer the expected (from an adult perspective) event over the unexpected, they clearly differentiate between normal and bizarre events. By seven months, the evidence is even stronger that babies know that objects need physical contact to move but that people can move on their own.[10] Further, around nine months old, children are generally starting to point to things to direct others' attention and watch others' eye gaze to discern what they are attending to. These two critical achievements in social interaction and social learning indicate an appreciation that agents can be prompted to act without having to contact them.[11]

Agents Act to Attain Goals: The Jumping Circles Experiment

Babies know that agents, unlike ordinary objects, can move themselves and can move other objects such as hats. They also know that this activity on the part of agents isn't willy-nilly but is done to accomplish goals, such as to look at something, go someplace, or get something. Again, clever experiments give us evidence.

By one year old, babies have a growing awareness of people as agents—able to move themselves around without being physically contacted in goal-directed ways. György Gergely and Gergely Csibra showed babies a computer-animated display in which a small circle "jumped" over a barrier and contacted a larger circle.[12] After babies grew tired of this display, they were shown one of two new displays without the barrier. Either they saw the same small circle move in a straight line and contact the larger circle (through the space where the barrier used to be), or they saw the small circle jumping over where the barrier used to be and contacting the larger circle—exactly the same action as before. Generally in these sorts of habituation experiments with infants, it is the unexpected or novel display that recovers babies' attention. In this experiment, only babies who saw the same jumping motion recovered their attention. Why? In this and several other experiments using different animated scenarios, twelve-month-olds appear to expect objects with an established goal (for instance, to get to the large circle) to continue to pursue that goal in the most direct and efficient manner available (that is, without jumping if there is no barrier in front of the large circle).[13] When movement was no longer motivated by the goal, babies acted surprised and looked longer.

Agents Need Not Resemble Humans: The Fuzz Ball with a Face

Though we sometimes take for granted our ability to distinguish agents from nonagents, this business can be tricky even for normally functioning humans. I once had a summer job that involved maintaining long-term felony records for the county district attorney's office. The storage site was a hot, dusty metal warehouse on the grounds of a small municipal airport where I retrieved and returned files and generally organized the facility. Paperwork and supporting materials from some of the county's most notorious crimes bulged from the brown banker's boxes, including the photographs of brutal murder scenes. I worked alone, alone except for the mannequins that were used as props for trials. Not infrequently these mannequins—even though I knew they were there and only mannequins—startled me or otherwise made me feel uneasy, as if someone were watching me. Human forms, especially human-like faces, are easily and rapidly detected by people from early infancy. A human-like appearance is one important sign of a potentially helpful (or dangerous) person being around. To function as normal people, however, we have to understand that looking like a human and being an intentional agent are not the same thing.

Additionally, we must understand that being an intentional agent does not always mean that the being is a human. If something had to look human to be an intentional agent, lions could literally walk into towns and drag away people because we would assume they were no more intentional and self-directing than dandelions. But we know that while looking like a human is one cue for a potential agent, it is not the only one or even the most reliable. Our ability to think about things that do not resemble humans as agents also allows us to conceive of wispy vapors as thinking spirits, mountains and volcanoes as wrathful judges, and elephant-headed four-armed gods as willful agents.

What about babies? In the jumping circles experiments, the researchers capitalized on the fact that children's reasoning about agents does not require the agent in question to remotely resemble a human. Susan Johnson and collaborators made the same point in a different sort of experiment with twelve-month-olds.[14] This study made use of the finding that by around nine months old, babies follow the gaze of other people and look to see what they are looking at.[15] With this information in mind, these twelve-month-olds were shown a furry object on a table. The object looked like a large fuzz ball with a smaller fuzz ball attached to the near side of it. For one group, it babbled whenever the baby babbled and flashed an internal light when the baby moved, as if it reacted to the baby. For another group, it behaved the same way but had a face (two eyes and a nose on the smaller fuzz ball). A third group of babies saw the object with a face make the same amount of noise and light flashing but independent of the baby's action and arbitrarily (not directed toward a goal). A final group of babies watched the faceless object make the same noise and light flashing arbitrarily. After the babies had watched the object for one minute, it made an attention-grabbing beep and then turned toward one of two targets on either edge of the space. The measure of interest was whether babies followed the "gaze" of the object, thereby suggesting that they understood it to be an agent with directed attention.

The babies were significantly more likely to follow the gaze of the first three objects, that is, if the furry object had a face or if it interacted with the child (flashing in response to movement or making noise in response to babbling). The contrasting fourth, faceless object babbling and flashing arbitrarily did not get this response. These results suggest that having only one of two properties (having a face or interacting in a purposeful way) is sufficient for children to assume that an object has some kind of attention—a rudimentary mental state.

Here again twelve-month-old babies demonstrated that they possess valuable knowledge about intentional beings or agents. Once again, the agent in question did not have to look at all like a human being. Experiments such as these suggest that from the first year of life, children's developing conceptual systems for making sense of agents are not specifically restricted to humans, leaving open the possibility that babies could use these same conceptual systems when thinking about gods.

Agents Need Not Be Visible

As all parents know, babies quickly learn to call and cry for us to come to them. That is, babies eventually understand that even when parents are not visible, they can act (to come to us) and be acted on (summoned through cries). To function properly in social groups, survive the threat of predators, or capture prey, we (like our ancestors) must be able to think about agents we cannot even see. Tracks, traces, noises, or even inexplicable silence need to be useful as cues that an agent might be around, watching and waiting. We need to be able to use this information usefully, for instance, to realize that the sight of the lion's prints pointed toward our home tells us just as much about the lion's goals as seeing the lion itself. If we were unable to reason about the minds of other people not visibly present, we would not be able to hold conversations in the dark or on the phone, we would not be able to anticipate future interactions and plan ahead, and the idea of government officials whom we have never seen making decisions and acting toward particular ends would be entirely opaque to us. Fortunately, we naturally develop the ability to think about unseen agents.

This achievement is just a matter of babies' putting together the other principles about agents mentioned above. Objects cannot move themselves, but agents can move them. If the furniture moved about during my nap, it is likely that my mother or father moved it—and not likely that the furniture moved itself. Babies learn the ability to reason about an agent, even when they did not see the agent perform the acts they have observed, and even when they did not see the agent at all. Maybe babies, too, can reason that other sorts of changes in the world—when the furniture of the world moves around—are the result of what agents are doing outside babies' direct observation. In the next two chapters, I present research that bears upon this speculation.

Perhaps the strongest evidence that young children can readily conceptualize and believe in the existence of nonvisible agents comes from research on children's invisible or imaginary friends. Upwards of 40 percent of us will have at least one imaginary friend during early childhood.[16] Even children with abnormal theory of mind development can have rich interaction with invisible companions. *Theory of mind* refers to the ability to reason about others' actions in terms of mental states such as beliefs and desires. Autism and Asperger's syndrome are characterized, in part, by theory of mind deficits. A friend's son with Asperger's syndrome was about four years old when I stopped in for a visit, and I found that his house was home to three invisible dogs, the most rambunctious curiously named Sin.[17] If children can have chatty tea parties with a host of invisible people, the invisibility of gods is no problem.

Bradley Wigger recently conducted a series of interviews and experiments with children who had invisible friends at the time of the study.[18] I describe the experiments he used in Chapter 4. His basic finding, however, is simple. Three- to five-year-olds typically attributed superknowledge to their invisible friends, reasoning about them as they did about God in the experiments.[19] These findings can be interpreted as showing that children spontaneously create their own invisible, superhuman agents, or *gods.*

To summarize the research presented so far, we have good reason to think that in their first year of life, babies already have a handle on the basic properties of physical objects and understand that agents can break these rules. Specifically, agents can move themselves and other things, act to attain goals (instead of just moving arbitrarily), need not resemble humans, and need not be visible for children to think about them and attribute actions to them. These achievements are building blocks for thinking about humans and animals, but also about gods. Gods are agents that are often unseen or not resembling humans but acting nonetheless. But having the timber for thinking about gods and any inclination to do so are two different things. In the next two chapters, I say much more about children's natural inclinations to think about gods, but I lay some groundwork in the remainder of this chapter by trying to convince you that we are not just capable of thinking about gods from the early days of life, but we have a natural propensity to eagerly look for agents in the world around us, whether or not they resemble the people or animals we know.

Eager Agent Detectors

When my daughter was a toddler, she would occasionally encounter an earthworm while she played in the yard. Instead of recoiling, she would pick it up and carry it around, cooing and talking to it as if it were an agent. (This sort of behavior convinced my wife and me that it was time to buy the child a doll.) I share this story to highlight that children do not merely possess the requisite mental tools to reason about agents when an obvious, indisputable candidate for agent arrives on the scene. Rather, they happily extend agency to objects that many adults would argue are not proper agents.

The body of psychological research I have described shows that children possess the necessary tools to think about agents, and also that children eagerly look for and think about agents. Children could use conservative criteria for treating something like an agent, unconsciously reasoning, "It doesn't follow the rules for ordinary objects, but I need to exhaust all other options before treating it like an agent." Children do not use this conservative strategy. Rather, they seem to have a disposition to reason about things as agents when they meet just the barest of qualifications, so that even red and blue dots could qualify. If it breaks the rules for ordinary objects (naive physics) and it seems to be goal directed, then it is treated as an agent until a reason is given to believe otherwise.

Attributing Agency to Nonhumans: Chasing Disks

Philippe Rochat and colleagues conducted a series of experiments looking at whether babies might be sensitive to agent-like movement that is not performed by a human.[20] In these experiments, the researchers presented three- and six-month-olds with two computer displays. Both monitors showed a blue disk and a red disk moving around, but in one display a disk was "chasing," but never "catching" or touching, the other. On the other monitor was a display that was identical to the first in terms of the average speed of disks, rate of direction changes, average distance between the disks, and so forth. The difference? One disk was not following the other but moving arbitrarily. If babies could tell the difference between these two displays, it would have to be because of the causal relationship between the movement of one disk and the movement of the other—a kind of relationship seen only in agents. Sure enough, both the three- and six-month-olds showed differences in which display they preferred to look at, indicating that they could tell the difference between them. Six-month-olds performed similarly to adults, watching the nonchase longer than the chase. Perhaps by then, they could already quickly figure out what was going on with the chase but (like an adult) wondered what could be happening in this other, causally unusual display.

By around age nine months, babies in a similar follow-up experiment showed that they were not just sensitive to the causal relationship between the two disks in the chase display; they could also tell who was chasing whom (so to speak). In this experiment, babies first watched either the red disk chasing the blue one or vice versa until they got habituated—good and bored. Then the experimenter reversed the chase. If the blue had been chasing the red, the red began chasing the blue. The eight- and nine-month-old group (but not the younger children) noticed the difference and started watching again (they "dishabituated").[21] Apparently they recognized the role reversal.

Not bad for someone who cannot walk, talk, or use a toilet, and notice that many of these experiments used animated displays of colored disks that do not remotely resemble a human being or an animal. Babies do not need a person or even a known animal present to get their agency reasoning running—an important point if they are going to apply their reasoning about agents to part-animal monsters or invisible gods. The system is easily turned on by action that violates normal object behavior: ordinary objects do not "chase," so they must be agents.

Circle and Triangle in Love

It appears too that this tendency to treat things as agents at the moment when ordinary object reasoning fails persists into adulthood. When computers do not respond to our button pushing, we often find ourselves yelling at them—but not just yelling anything. We curse the computer for its mental failings, its sinister intentions, and its antihuman motives. We treat it as an agent. Similarly, a fisherman would not say, "The movement of that lure in the water excites the retina of the fish in such a way that a signal is sent to its brain, activating its prey pursuit instinct." But I have heard many fishermen describe in details the sophisticated thought patterns of elusive fish, their fears and wants: "That bass thinks the lure is a frog and he just loves little froggies!"

Children are not the only ones ready to treat geometric shapes on a screen as agents. In a now-classic set of experiments from the 1940s, Heider and Simmel showed college students a film of a circle and two triangles moving around and in and out of a broken square.[22] Then the experimenters asked the students to "write down what happened in the picture."

Remarkably the young adults described the shapes as having rich mental states including beliefs, wants, aims, intentions, and even gender roles. Heider and Simmel reported that only one of thirty-four adults in their first study avoided describing the scene as if the shapes were people. Here is an unusually elaborate example of the sort of attributions made to these two-dimensional geometric shapes:

> Triangle number-one shuts his door (or should we say line) and the two innocent young things walk in. Lovers in the two-dimensional world, no doubt; little triangle number-two and sweet circle. Triangle-one (hereafter known as the villain) spies the young love. Ah! . . . He opens his door, walks out to see our hero and his sweet. But our hero does not like the interruption (we regret that our actual knowledge of what went on at this particular moment is slightly hazy, I believe we didn't get the exact conversation), he attacks triangle-one rather vigorously (maybe the big bully said some bad word).[23]

As fanciful as this description might sound, this sort of experimental finding has been replicated many times over using objects on video displays that do not begin to resemble humans or even animals, but rather look more like ordinary inanimate objects.[24] In my own work with Amanda Johnson, we asked college students to narrate their own actions while placing ball bearings over holes on a board.[25] Periodically an electromagnet sent the ball bearings racing toward two different spots on the board in violation of naive physical expectations for ordinary objects. The result? Almost two-thirds of participants spontaneously referred to the ball bearings as if they were agents. Students made comments such as, ". . . and a couple ones did not like me," "That one did not want to stay," "Oh, look. Those two kissed," "They are kind of fighting," and "They are not cooperating."

This tendency to treat various objects as if they were intentional agents is so common that we often do not even recognize it for what it is. A certain kind of causal relationship between objects—when they violate regular mechanistic causation in a way that looks goal directed—sets our minds racing to interpret the actions as if the actors are human-like agents, whether or not they look like people or animals. I do not emphasize this very human trait to make fun of it or to suggest that people are basically error prone when it comes to recognizing agents. Determining the accuracy of our mental equipment for finding agents and their actions is not easily done, but I suspect that this equipment is actually fairly accurate: when we walk into a room full of agents such as people and family pets, we pick them out immediately and ignore the lamps, chairs, and televisions. I highlight the "mistakes" (if they are mistakes) to illustrate just how eager we are to apply our ability to reason about agents whenever and wherever it seems to do the job.

Spotting Spirits

Occasionally we have experiences that we attribute to unusual agents. Here is a true story. Several young girls slept in the living room of the house where their great-grandfather's clock sat on a shelf dutifully keeping time. In the middle of the night, one of the girls awoke the others, exclaiming something about a ghost coming out of the clock. "I don't see anything." "I swear, it came right out of the clock!" "Oh, don't be such a chicken. Go back to sleep." "I'm not a shicken." In the morning, the clock was surrounded by tiny wooden crosses placed as a precaution against return visits.

This tendency to identify things—objects, shadows, strange lights—as agents, even given only slight evidence, is just how our agency detection system works. This brain function is sometimes referred to as the *hypersensitive agency detection device* (HADD), to emphasize that sometimes it seems too eager to find agents. Whether it is hypersensitive as opposed to just sensitive enough is uncertain, but it certainly is not bothered by the agent or agency in question looking very unlike a human.

Anthropologist Steward Guthrie has argued that there would be very good evolutionary reasons for our agency detection faculty to err in the direction of overdetection: better to be safe than sorry.[26] Occasionally thinking of the wind in the brush as a some*one*—an agent—usually is not very dangerous or costly to your survival. If you miss an agent in your environment, however, you might be missing a potential friend or foe, predator or prey. You could become tiger chow. Guthrie suggests that the tendency of this device to easily find decidedly nonhuman agents in our environments may be one important genesis or propagator of beliefs in gods. Certainly the ability of the HADD to spot certain happenings as indicative of even nonhuman agents reinforces beliefs in gods. If our HADD found only decidedly human agency, we would never think that storms and plagues were signs from the gods or that strange noises and happenings were messages from the dead. We do, thanks to HADD.

Children are born with minds ready to make sense of the world around them. Because we humans are not particularly strong, fast, agile, or tough for their size, our big advantage is intelligence. We have a tremendous ability to find meaning and relationships among all that we experience. We quickly learn to anticipate, to wonder "what if?" and to figure out the causes of events. Babies have minds specially equipped to solve these problems.

Babies also need to be able to identify agents because they represent both our greatest threat and our promise for survival and fulfillment. We humans rely on each other to survive and thrive, so we must make sense of each other. But we also need to understand animals. Unlike rocks or tables, animals get up and move in purposeful ways. Some appear to think and plan. If we cannot outsmart potential prey, we could miss a much-needed meal. If we cannot anticipate the actions of a dangerous predator, we could become a meal for someone else.

So it is little surprising that infants show signs of knowing the difference between agents and inanimate objects and that they show a great sensitivity to the possible presence of agents around them.

We are on the lookout for agents that are nonhuman, unseen, and acting on the world when we are not looking—secret agents. These abilities make it very easy to think about gods. With these abilities in place in infancy and with a heightened tendency to detect agents or attribute agency in the face of little encouragement, very young children are receptive to the idea of gods; it may even be that were children not provided with ideas about gods, they would discover gods for themselves when combined with a tendency I explore in the next chapter—that of finding design and purpose in the natural world.

Two

Children in Search of a Purpose

The title character Ariel from Disney's animated movie *The Little Mermaid* is fascinated with everything in the human world and has a vast collection of human-made things she has found in shipwrecks and on the seabed. Ariel's chief source for insights into what things are and what they are for is Scuttle, a floppy, pontificating seagull. In one scene Ariel brings Scuttle a bag full of "human stuff" that she has collected. Scuttle oohs and ahhs as Ariel produces a fork. Scuttle explains that this is a dinglehopper. "Humans use these little babies to straighten their hair out." When Ariel hands Scuttle a pipe, he squints and says it is a very rare snarfblaster, a fine musical instrument humans use to entertain one another. This Scuttle-like compulsion to figure out what stuff is for, that is, to determine the function and purpose of objects, is natural to children. Not only do kids look for purpose in human-made things (artifacts) like forks and pipes, but also in natural objects like rocks and rivers, plants and animals.

In the previous chapter, I shared how babies understand agents as distinct from ordinary physical objects. Agents can move themselves; ordinary physical objects cannot. Agents have beliefs, desires, and intentions that motivate their actions; ordinary objects do not. Very young children successfully think about agents that are not visible, and agents need not physically or otherwise resemble humans. This ability to reason about agents enables even young children to think about gods—typically unseen nonhuman agents.

Even more than having the ability to understand agents, children have a tendency to identify objects as agents and see events as caused by agents with little provocation. This tendency to easily find agents (sometimes without large amounts of evidence) persists into adulthood and makes the discovery of gods not only possible but likely. Still another feature of the young child's mind also contributes to a readiness to embrace the existence of gods: kids eagerly detect design and purpose in the natural world.

What's That For?

Do you ever wonder why certain plants or animals have the features they have? As long as I can remember, I have wondered what things are for. When I see the proboscis nose of the elephant seal, I wonder, "What is that for?" Or the spiky covering of the horse chestnut. What is that for? Why do zebra have stripes, jackrabbits have enormous ears, and kelp plants have air pouches at the base of their leaves? Maybe I am unusually curious, but I am not alone in being drawn to questions about why things are the way they are in the natural world. When encountering a strange animal at a zoo or while watching television, I have heard other adults wonder at the strange features of plants and animals or delight in discovering what a thing is for.

This wondering about what a thing's purpose or function might be starts early in childhood. I have heard children, my own included, wonder about what elephant trunks are for, and they delight in discovering the answer. My daughter asked why roses have thorns, and an account of how the genetics or biochemical processes within the rose plant gave rise to the thorns was not what she wanted to know. She wanted to know what function the rose thorns served, what their purpose was. This kind of reasoning about purposes, design, and function is sometimes termed *teleological reasoning.*

Perhaps a difference between children and adults, however, is the range of things to which they attribute purposes. People of all ages might expect that parts of animals or plants have purposes and that machines, tools, and other human-made things do too. Children's teleological reasoning extends to nonliving things in nature such as rocks, snowflakes, and mountains and also to whole animals and plants. For the child, it is perfectly reasonable not only to wonder what an elephant's trunk and ears are for, but also what the elephant is for, what a river is for, or what a rock outcropping is for. Developmental psychologist Deborah Kelemen has termed this eagerness *promiscuous teleol-*

ogy—teleology meaning that children find design and purpose in the world, *promiscuous* in the sense that they find purpose with very little evidence and in places that adults might find inappropriate.

In one experiment investigating promiscuous teleology, Kelemen presented children with a disagreement between two fictitious characters, Ben and Jane.[1] One argued that one of a number of things was made *for* something. The other character disagreed. The four- to five-year-old children were asked to decide who was right, Ben or Jane. The items under consideration were living things (woman, man, baby, tiger, cat, binturong, tree), biological parts (human earlobe, lion leg), natural objects (cloud, iceberg), natural object part (mountain protuberance), artifacts (jeans, ring, tryogaster), and an artifact part (clock hand). Children were shown a photograph of each object and pictures of Ben and Jane. An example of an item appears below:

> See this. This is a tiger.
>
> Ben says a tiger is made for something. It could be that it's made for eating and walking and being seen at the zoo or it could be that it's made for other things. But Ben is sure that a tiger is made for something and that's why it's here. Jane says that is silly. A tiger isn't made for anything. Even though it can eat and walk and be seen at the zoo, that's not what it's made for, they're just things it can do or people can do with it. Jane is sure that a tiger can do many things but they aren't what it's made for and they aren't why it's here.

> Point to who you think is right. Ben who thinks a tiger is made for something or Jane who thinks that's silly because a tiger isn't made for anything.[2]

Perhaps surprisingly, these four- to five-year-old American children were just as likely to agree that the natural objects and living things were made for a purpose as the human-made objects. Three-quarters of their answers favored a functional, "made-for" answer. Given the nature of these questions and examples, it is a stretch to argue that children were simply taught these kinds of explanations for what tigers are for. Surely American parents do not teach that tigers and trees are made for something with the same conviction or frequency with teaching that machines and jeans are made for something. Children seemed very drawn to these answers that suggested a purpose or function for the natural things' existence.

In another set of experiments, Kelemen found that seven- and eight-year-olds preferred teleological (functional) explanations for the features of various animals and natural nonliving objects.[3] For instance, when shown a picture of a pointy rock and asked, "Why do you think the rocks were so pointy?" children favored purposeful, functional explanations such as, "They were pointy so that animals wouldn't sit on them and smash them," over physical explanations such as, "They were pointy because bits of stuff piled up on top of one another for a long time." Again, parents and teachers do not teach kids that rocks are pointy so that animals will not sit on them. This is one way we can be confident that these tendencies are not merely a reflection of indoctrination on the part of adults.

Further, Kelemen and collaborator Cara DiYanni asked six-, seven-, nine-, and ten-year-old British children about the origins of a number of natural and artificial things.[4] These things were divided into natural events (thunderstorm and flood), nonliving natural objects (river and mountain), animals (monkey and bird), and artifacts (boat and hat). They first asked children open-ended questions in the form, "Why did the first-ever flood occur?" and "Why did the first-ever bird exist?" In keeping with their tendency to find purposefulness in the natural world, these children offered teleological answers more than any other type of answer for natural objects, animals, and artifacts. Only for natural events did they favor physical explanations (for instance, the first thunderstorm occurred because "two clouds bashed together"). To illustrate the sorts of answers children offered, consider these functional explanations from children in the study:

Why did the first ever bird exist?
"To make nice music"
"Because it makes the world look nice"

Why did the first monkey exist?
"So then we had somebody to climb trees"
"So there can be an animal in the jungle"

Why did the first mountain exist?
"They made mountains . . . so people can look at them then maybe can get a piece of paper and draw"
"Cos the earth, perhaps, there were usually always—all the time—earthquakes and they needed something

to . . . like a paperweight . . . like perhaps there was lots of earthquakes so they thought that there should be something that could stop them so they put lots and lots of weights of stone"

Why did the first river exist?

"So boats could come in the water"
"So that people could do fishing"[5]

After children answered these open-ended questions, the interviewer asked closed-ended questions. The questions of particular interest here were the so-called intelligent design questions. For these questions, the interviewer said, "So we've been talking about mountains [or rivers, monkeys, birds]. Now here's the question. Did someone or something make the first ever mountain exist or did it just happen?" Children could answer one of three ways: they could say that some*one* made it, some*thing* made it, or that it just happened.[6] For natural events and natural objects, about half of the children said that some*one* made it, and about three-quarters chose some*one* answers for animals. In the open-ended questions, children showed a tendency to see natural objects as having a functional purpose. In these closed-ended questions, children exhibited sympathy for someone—an agent—being the cause of that purpose. Features of the natural world, then, are not just accidents that happen to be useful, but are ordered and purposeful.

Only Agents Create Order

If you have spent much time hiking in wilderness areas that people regularly use for backpacking and camping, you may have discovered seemingly orderly arrangements of objects even far from civilization. Perhaps you have come across stones placed in the shape of rings, often with charred wood remains inside the ring. Or you have seen small piles of stones on the forest floor next to a well-trodden path. Maybe you have even discovered fragments of trees that have undergone some kind of process by which their bark has been removed and they are rendered relatively flat and this remnant then has symbols etched into it (signs). Instead of pondering just which set of natural events led to these seemingly ordered arrangements, you undoubtedly took these arrangements as signs that human beings had been there. After all, agents bring about purposeful order; natural events do not. Adults know this. But do babies? Recent evidence suggests that by twelve or thirteen months of age, they just might.

Developmental psychologists George Newman and collaborators presented twelve- to thirteen-month-old babies with two computer-animated displays: a ball knocking over a neat stack of blocks (obscured by a barrier for the actual striking) into a disordered heap, and vice versa (the blocks starting in a disordered heap and finishing in a neat stack).[7] Adults would immediately see that something unexpected had just happened: balls cannot stack blocks! But did babies show surprise? Using relative time spent staring at the display as a measure of jaw-dropping amazement, babies were indeed surprised. They looked longer at the surprising "ordering test."[8] This suggests that babies find a mere ball creating order more surprising than a ball creating disorder. More interesting still was Newman's second version of the experiment.

In this second version, a ball-shaped object with a simple face on it that scooted along like an agent (instead of rolling) disappeared behind the barrier and either (apparently) ordered or disordered the blocks.[9] In this case, the babies showed no preference between the ordering test and the disordering test. That is, they seemed to find neither display more surprising than the other. The most straightforward explanation of these and other similar experiments is that babies of around one year of age have the same intuitions adults have: people, animals, gods, or other agents can create order or disorder, but nonagents, such as storms or rolling balls, create disorder.

Given this sort of evidence, Kelemen and colleagues' findings that young children think some*one* accounts for what they see as purposeful order in the world is not surprising. From early on, children associate intentional agents with order; they see purpose, function, and design in the natural world; and so they naturally assume someone brought it about.[10]

Existential Theory of Mind

In July 2003 a visiting evangelist at an Ohio church asked God in a rhetorical flurry what God wanted to say to the congregation. Moments later, the church's steeple was struck by lightning, setting the church on fire. One witness from the congregation was quoted as saying, "He was asking for a sign and he got one."[11] Not surprisingly, this event caught the attention of the congregation and the popular press. The event seemed more than mere chance. Developmental and evolutionary psychologist Jesse Bering has suggested another way in which from childhood people possess a bias to see purpose in the natural world. He calls it *existential theory of mind.* Behind this imposing term is a simple idea: people have a strong tendency to search out and postulate meaning behind events. For instance, a friend once shared with me his experience of being outside under a starry night sky struggling with whether God was real and mattered to his life. He saw a brilliant shooting star blaze across the sky, which none of his fellow stargazers managed to see, and he immediately took the shooting star as a sign from God. My friend understood the event to mean something—to indicate something and be caused by some*one.* Bering argues that this search for meaning typically involves the assumption that the intentions of some agent lurk behind striking events. When calamity befalls us under unusual or surprising circumstances, we sometimes wonder, "Why me? What does it all mean?"

The ability to see events as symbolically communicating something appears to be a psychological achievement arising in childhood. Bering has conducted laboratory experiments with children exploring their ability to recognize events as potentially meaningful communicative acts.[12] In one of his "Princess Alice" experiments, the procedure went like this: Children are brought into a room and seated at a table. Two boxes sit on the table. The experimenter explains that one of the boxes contains a prize. If the child chooses the right box, he or she will win the prize. The experimenter also explains that an invisible princess named "Princess Alice" likes to help good little children and just might try to help them win the prize by somehow letting them know if they've made a wrong choice. A picture of Princess Alice (looking suspiciously like a Barbie doll) hangs on the wall. The room also has a table lamp in it glowing brightly. After this explanation to the child, the experimenter asks the child to choose a box but then makes an excuse to leave the room immediately with the child's parent for a moment. When the experimenter returns, they will open whichever box the child has her or his hand upon. The child reaches for a box and then, as she or he touches a box, either the lamp flickers or the picture of Princess Alice falls from the wall. The measure of interest was whether children change their choice and, if they do, how they explained their choice.

Bering's results showed that seven-year-olds reliably changed their decision under these conditions and reached for the other box (but not when the lamp did not flicker or the picture did not fall). They also explained that the flickering light or the falling picture was a sign to them to change their choice. Bering also reports that the ability to use events as meaningful signs appears before the ability to explain. Younger children changed their choice when the picture fell off the wall but could not explain why.

Research in this area is still new, but the idea that people have a fairly early-developing predilection to endow events with meaning and intention is a provocative idea that, if borne out by more systematic evidence, would contribute to our understanding of religious beliefs. Not only might children have a tendency to see the natural world as purposeful and designed, but they might also regard additional events such as thunderclaps and shooting stars as having communicative purposes, at least under some conditions.

Adults Finding Purpose

Importantly, this tendency to see intelligent design does not appear to be something we simply outgrow. Kelemen has recently demonstrated experimentally that adults, even scientifically trained ones, possess a bias to favor purpose-based explanations.

Kelemen and Rosset asked college students to evaluate various explanations for natural things and their particular features (such as the white fur on a polar bear).[13] Among the explanations were scientifically "good" explanations (not invoking design or purpose but only natural processes) and other explanations that included both purpose-based explanations and just strange ones (on the order of polar bears being white because the sun bleaches their fur). When they had plenty of time to consider their answers, adults often rejected that teleological explanations were "good" explanations, just as they had been taught in their college science classes. Evolution has no purpose, direction, or design, we are taught, so such explanations were often classified as "bad" (along with sun-bleached polar bears). But when Kelemen forced them to answer quickly, these students began accepting the teleological explanations as "good" while their accuracy in (correctly) rejecting other "bad" explanations actually improved (possibly because they did not

Children come into the world ready to understand willful intentional agents, which makes them able to conceive of gods, but does not necessarily motivate them to think about gods. As long as the world around us concerns other humans, human achievements, and human activities, why bother with gods? The simple answer is that, like the weather, much of life is out of human control.

By adulthood, most reflective people quickly realize that the world is not only about humans. The origins of the universe, the laws of nature, the diversity of plants, the whims of the weather, success at raising crops or finding enough food, and many other matters fall outside full human control. Attributing the cause (or blame) for a great many events or conditions to gods or a God strikes many adults as reasonable. But do babies have the same tendency? Do young children look beyond human activity for the causes of what they see around them? Increasingly, evidence from developmental psychology suggests that the answer is yes.

As discussed in the previous chapter, children readily attribute purposes to many natural phenomena, including the features of animals, plants, and even rock formations. Such purposefulness all around might prompt children to embrace the existence of a god or gods behind it all—a designer to account for the design. But just because it seems a sensible connection for many adults does not mean children make the same connection. After all, perhaps it is a very different sort of willful being, or beings, that account for the perceived order and purpose of the natural world—maybe even people.

Children: Intelligent Design Theorists

Swiss psychologist Jean Piaget, easily the most influential developmental psychologist of the twentieth century, argued that children naturally and spontaneously see the world as human-made. Based on interviews with children, Piaget concluded that until age eight or so, children have a strong tendency to believe that the natural world was made or manufactured. He called this *artificialism*: the idea that rivers, trees, celestial, and animal bodies are artifacts made by someone just as much as furniture, tools, and toys are artifacts made by someone. Mountains, lakes, the sun, and the moon have not always been, but at some point were made. Such a claim is consistent with Kelemen's finding that children are prone to see elements of the natural world as intentionally and purposefully designed, but Piaget went a step further, determining whom children identify as the maker of mountains, lakes, and the moon. Human beings, according to Piaget, are the creators in the minds of children. Consider the following exchange with a six-year-old that Piaget reported. "Where does the sun come from?—*From the mountain.*—How did it begin?—*With fire.*—And how did the fire begin?—*With matches.*—And how did the mountain begin?—*With the earth. . . . It was people who made it.*"[1]

Piaget argued that this tendency to think of the natural world as being made is not simply due to confused religious education. "One finds . . . that before any religious teaching has taken place, artificialist questions are being framed by children of 2 to 3 years old. 'Who made the sun?' asked Fran at 2 years 9 months."[2]

Piaget's account of artificialism does allow a role for God or gods in the process. Children explain that God was one of the people, one of the engineers, or perhaps the foreman of the construction site:

> On one hand, the child is taught that a God has created Heaven and Earth, that all things are directed by Him and that He watches us from Heaven where He dwells. There is nothing surprising in the children simply continuing to think along the same line and imagining in detail the manner of this creation and supposing that God secured the help of a band of skilled workmen.[3]

The reason such an easy incorporation of people into God's creative activity was that from Piaget's perspective, children could not distinguish between God and humans until age eight or nine. God was a man (or humans were gods):

> Moreover, when the child speaks of God . . . it is a man they picture. God is "a man who works for his master" (Don), "a man who works to earn his living," a workman "who digs," etc. In short, God is either a man like other men, or else the child is always romancing when he speaks of him, in the same way that he speaks of Father Christmas and the fairies.[4]
>
> . . .
>
> He is just a man like anyone else, who lives in the clouds or in the sky, but who, with this exception, is no different from the rest. "A person who works for his master," "A man who earns wages," these are of the type of

definition that working-class children of about 7–8 give of God.[5]

If "God" meant just a particular human being, then any children's reports of God or people having made a natural object were evidence of artificialism. It did not matter whether children distinguished between God and people when it came to creating. Piaget wrote, "Certain children, for example, have hesitated in attributing the lake to God or to men, saying: 'I don't know if it was God or some men who did it.'"[6] Likewise, he said, "As we have seen, even the sun and the moon and the sky are attributed to the activity of man and not of God, in at least half the cases."[7]

Based on Piaget's work in the early twentieth century, then, children regard the natural world as the product of human creative activity. Humans, perhaps including a special bloke named "God," went about making lakes, mountains, the moon, sun, planets, and sky a long time ago just as they build houses, levies, dams, and bridges today. Such an explanation is certainly compatible with more contemporary research regarding children's natural inclination to see the natural world being purposefully designed.

To children as to adults, it seems that design intuitively implies a designer, since infants apparently expect agents, and not nonagents, to be able to bring order out of chaos.[8] Additionally, in their studies of children's ideas about origins, Deborah Kelemen and Cara DiYanni found that children who offered purposeful explanations for why the first thunderstorm, river, birds, and other natural things existed tended to affirm intelligent design answers to a later forced-choice question. That is, for instance, children who answered the question, "Why did the first ever mountain exist?" with a purpose-driven explanation such as, "Because people could climb it," were more likely to go on to say that some*one* made the first-ever mountain exist (as opposed to saying some*thing* made it exist or that it just happened). These findings show a relationship between seeing design and assuming a designer.

From Piaget's perspective, then, humans have the ingenuity and industry to fill the role of designer. Naturally, then, children see the world as intelligently designed by humans. Such an interpretation fits many of the facts and has a certain ring of plausibility. After all, it is not implausible that from the perspective of the newborn, superpowerful beings (parents) manage their world: when it is light or dark, when I get food or not, when I move from place to place, and what is around me. Agents do account for much of the design, order, and purpose that a newborn experiences. Perhaps young children extrapolate from these experiences that the same is true of the rest of the world: adults are powerful people who have designed and built the world. Piaget's position, however, has not been without challenges.

Humans Do Not Make Natural Things

One concern about Piaget's account was the sort of interviewing technique he used to get opinions from children. Perhaps asking them where things came from in the way he did inadvertently encouraged children to invent origin stories even when they did not really have any particular belief. Perhaps describing people building mountains and lakes was the easiest account for them to think of and explain, but they did not really think about it until they were asked the question. A delightful thing about children is their jazz-master ability to improvise and invent explanations and stories on the fly about things to which they have never given much thought before.

To avoid these potential problems, other developmental psychologists have more recently used different techniques to address whether children understand that people make tools, machines, and other artifacts but not natural objects and substances. Susan Gelman, for instance, simply asked American children whether a number of different items were made by people: "Do you think people make lemons?" This yes/no question did not require children to have to explain the origins of the items; if they just guessed, we would expect that their rate of producing "yes" answers should be the same as "no" answers and similar for different items. In this task, 80 percent of the time the four-year-old participants accurately answered that people make artifacts such as dolls and telephones but not natural objects such as lemons or birds.[9]

In a subsequent study, Gelman and collaborator Kathleen Kremer asked children specifically about the sorts of natural objects that Piaget did: the moon, sun, stars, ocean, clouds, and thunder.[10] They first presented each child with a picture of the object (except in the case of thunder), asked the child to identify it, and then asked, "Do you think people made it?" In contrast with Piaget's reports, less than 30 percent of four-year-olds and 10 percent of seven-year-olds said people made these natural things. When asked about whether people made various artifacts (cups, dolls, shoes, cars, hammers, and televisions), children in both age groups were essentially perfect in saying that humans make them.[11]

Similarly, Olivera Petrovich presented British preschoolers with pairs of photographs and asked whether the pair included something that could be made by people.[12] These pairs included animals (dog), plants (daffodil), natural things (snow and leaves), common artifacts (chair or books), and toy animals and plants. When the pairs included a natural thing such as a leaf, in contrast with an artifact such as a bus, preschool-aged children nearly always accurately answered which one of the pair could be made by people. Only when one of the pair was a human-made replica (such as a toy cow) were children confused. On the basis of these results and other evidence (including Gelman's), Petrovich concluded that preschoolers clearly discriminate between the natural world and the artificial when considering where things come from.

Contemporary research, then, casts doubt on the inevitability that children think of the natural world as created by humans. Not by eight or nine years but by four years old (or younger), they understand that people do not make natural things, only artifacts. If the natural world appears so purposefully designed, then who did the designing if not humans? God appears to be a strong candidate for children to embrace in this designing role.

Here, again, Petrovich's research is instructive. She asked British preschoolers questions regarding the origins of planets, animals, and natural things such as the sky, earth, and large rocks—the same sorts of items that Piaget asked about.[13] Instead of using open-ended questions, Petrovich gave children three options: it was made by people, by God, or no one knows. Children in these studies were nearly seven times more likely to answer that God made natural things than people did. Perhaps unsurprisingly, British children understand that God is a better candidate for a creator than are humans.

Similarly, Kelemen and DiYanni asked six- and seven-year-old British children about the origins of natural and artificial things: natural events (thunderstorm and flood), nonliving natural objects (river and mountain), animals (monkey and bird), and artifacts (boat and hat).[14] The experimenter asked children open-ended questions in the form, "Why did the first-ever thunderstorm occur?" and "Why did the first-ever monkey exist?" For children who offered that some*one* instead of some*thing* was the reason, God was easily a more sensible "someone" than a human or humans. The someone was two times more likely to be God than a human if the question was about a natural object (river or mountain), two and a half times more likely to be God if the question was about a natural event (thunderstorm or flood), and twenty-five times more likely to be God if the question was about an animal. In contrast, if the question regarded an artifact (boat or hat), a human cause was almost twenty times more likely than a divine one. On subsequent forced-choice questions, children who had suggested some*one* was the reason favored God, Jesus, or Allah as the designer over 80 percent of the time for natural events, natural objects, and animals, but favored a human intelligent designer for artifacts 82 percent of the time. As with previous research, these children seemed to understand that gods make natural things and people do not, but people make artifacts and gods do not.

These results cast doubt not only on the claim that children see elements of the natural world as created by humans (Piaget's artificialism) but also the claim that children before the age of eight or nine cannot distinguish between God and humans. If I wanted to argue that to children, "God" just means a special human or a "man who works to earn his living," then I would have to accept the cumbersome idea that this God fellow is a peculiar sort of human: he is the only one who makes natural things but does not make artifacts. That sure sounds like a different sort of being than a human. Such insistence that God is indistinguishable from human beings for young children also fails to square easily with the experiments I share in the next two chapters.

Prone to Creationism, Not Evolution

An Oxford University colleague of mine tried to teach his adorable, curly-haired four-year-old the basics of evolution. He explained that long, long ago, our ancestors did not look as we do now but may have resembled something like an ape, and before that something like a lemur. The boy's deep brown eyes grew perplexed. "He looked at me puzzled while I explained the basics [of evolution] and emphatically said that it was not true, that we had always been like this (human)," reported the dad. At the offer that God made us this way, the boy returned to ease. Yes, that made sense, but "he had said 'no' when I suggested 'Santa Claus.'"

From infancy, children understand that agents can create order but nonagents cannot. By the preschool years (at least) children see things in the natural world as designed and purposeful. By four years old, children appear to understand that the designer of the world is not human. Given these predilections in childhood thought, suppose a typical child hears about a nonhuman intentional being named "God" that is responsible for designing the world and the things in it. Would this typical child find this "God" idea attractive? It seems so. Most kids would find such a notion very sensible sounding. It would come as no surprise that if children hear about a powerful God (or gods) that might have created the world, they would be ready to believe it. Intuitively, it just makes sense. A god explains the order and purpose they see. The way their minds develop makes them receptive to such an idea.

But what then happens if someone suggests that the apparent design of the natural world is not deliberately caused at all but is illusion? What if a child is presented with the idea that the "purpose" they see is not really purpose at all but the result of nonintentional, unguided natural processes, or, in the case of the cosmos generally, it may just be an unexplained happening? Because this idea flies in the face of their intuitions, most children (and perhaps most adults) will find such an idea odd, challenging to understand, and difficult to accept.

Anna is a beautiful, curious five-year-old girl. The daughter of Frederick, a Danish colleague of mine, Anna is like many other girls her age: she loves to ask all sorts of questions about life and the world—questions like *Why does it rain? What are elephants for?* or *Why don't I have a sister?* Yet Frederick and Anneke, her parents, do not fill her head with religious answers to these questions. They are proudly secular, well-educated urban Danes and put no stock in that religious tripe. They try to stick to the science—the physics, chemistry, and biology—of such things and give purely secular, materialistic explanations.

But even in publicly secular Denmark, children can bump into strange ideas about gods and the like and can come home with questions about these beings.

Frederick relayed to me that one day, seemingly out of nowhere, Anna asked Anneke if God had created the world. Anneke deferred this one to Frederick, who has special expertise in world religions and theologies. In typical Frederick fashion, he answered, “The world wasn’t created. It has always been here.” That seemed puzzling to Anna. She shook her head and said, “No. That couldn’t be right.” Frederick regrouped and explained, “Well, a long, long time ago there was this big bang and suddenly everything just appeared.” His daughter thought about this proposal and then commented, “God must have been surprised.” Naturally this answer surprised Frederick, but it does not surprise me and probably would not surprise many developmental psychologists who study how children learn about science. Many of the ideas science teaches us have to swim against the current of our natural ways of thinking. In this case, the apparently regular, orderly, and purposeful world around us, including the sun, moon, stars, and planets in their courses, are explained as nothing more than an uncaused appearance. Everything just happened to appear. Such a notion is surprising for most children and, from Anna’s perspective, would have surprised God too.

You may have noticed that in the studies that I have described, the most designed-looking part of the natural world was typically animals. Animals and their parts look designed and purposeful to children. So if children are told that this design in particular is illusion and has nothing to do with purpose or a designer, what would children likely think about such an account? Probably not much. Of course, this is exactly what children are taught about evolution by natural selection. Natural selection seems strange and implausible on its face to most children. Not surprising, then, developmental research has shown that children are resistant to alternative accounts for the design of natural things. In fact, the evidence from teachers and researchers is so strong that children have great difficulties understanding evolution that the Cognitive Development Society, a professional association of cognitive developmental scientists, hosted a special session on how to effectively teach evolution by natural selection during its 2009 biennial conference.

For instance, in a fascinating set of interviews with American children and their parents, psychologist Margaret Evans examined beliefs about the origins of many natural and nonnatural objects, including humans and animals.[15] Particularly, Evans compared Christian fundamentalists (defined as participating in a schooling that teaches a "literal interpretation of the bible") with geographically and educationally similar nonfundamentalist children and their parents. Not surprisingly, the fundamentalists' children more frequently offered creationist explanations of where animals came from than did the nonfundamentalists. After all, instruction from parents and teachers does matter for shaping explicit beliefs. Surprisingly, however, nonfundamentalist children showed a greater affinity for creationist accounts than their parents did, suggesting that the testimony of parents is not the only factor at play here.

Evans and her team interviewed three age groups of children and many of their parents. The three age groups were kindergarten through second graders (five- to seven-year-olds), third and fourth graders (eight- to ten-year-olds), and fifth through seventh graders (ten- to thirteen-year-olds). All children used comparable science and natural history curricula in school, with both fundamentalist and nonfundamentalist children even using the same textbooks in many cases. The key difference was that the fundamentalist children were in schools that also taught a literal interpretation of the Bible. Evans's team asked children both open-ended and closed-ended questions about where animals and humans came from. The open-ended question was phrased this way:

> I am going to ask you some questions. There are no right or wrong answers to these questions, just different kinds of ideas. Think about how the very first things got here on earth. A long, long time ago there were no things on earth. Then there were the very first things ever. Now think about the sun bear [a kind of bear from Asia]. How do you think that the very first sun bear got here on earth?[16]

In addition to the sun bear, children were asked about the tuatara, described as "a kind of lizard from New Zealand," and a human being. Evans selected two unfamiliar animals so that children would have to draw on their general ideas about animal origins instead of particular cases they may have discussed in class or at home. The answers to these questions were then coded for the type of explanation children offered. Did they say God or a person planned and put the animals on earth, did they say that the animal changed from another type of animal, or did they explain that the animal simply appeared or "grew from the earth" or some other such spontaneous generation explanation?

The close-ended questions concerned the same ideas but focused children in on a few possible origins of the first-ever thing. The interviewers again asked, "How did the first-ever sun bear get here on earth?" but then children had to rate their agreement with several different sorts of choices. The creationist option was, "God made it and put it on earth," and children had to say whether they disagreed a lot, disagreed a little, agreed a little, or agreed a lot. Likewise children rated an evolutionary option, "It changed from a different kind of animal that used to live here on earth," and either of two spontaneous generation options: "It just appeared" or "It came out of the ground."

Parents who participated in the study were asked slightly different questions. Instead of asking parents about their own beliefs in the origins of animals and people, the interviewers asked parents: "Imagine you were teaching a 10- to 12-year-old child you know, and he or she had asked you the following question. Please write down your answers. The questions were 'How did the very first sun bear [a kind of bear from Asia] get here on earth?'" and so on.[17] Answers to this question gave an indication of whether children simply answer the way their parents would tell them.

Interestingly, children's answers on both the closed-ended and open-ended questions differed importantly from what their parents said they would teach and what the school's science curricula presumably include. For instance, when considering open-ended answers, the five- to seven-year-old nonfundamentalists offered more spontaneous generation answers ("It came out of the ground") and more creationist ones than evolutionary accounts, even though their parents never said they would use a spontaneous generation account (and we know that the textbooks do not teach one either). Likewise, the eight- to ten-year-old nonfundamentalists frequently offered creationist accounts for where the first tuatara, sun bear, or human came from—more frequently than the rate at which parents said they would give creationist accounts. For the nonfundamentalist kids, creationist accounts appeared vastly more often than evolutionist or spontaneous generation accounts put together, even though these children's parents (as a group) reported equal likelihood to give a creationist or evolutionary account to a child. Only the oldest children from both backgrounds gave answers resembling their parents'.

Across fundamentalists and nonfundamentalists when considering the closed-ended questions and their agreement ratings, a similar pattern emerged. Creationist accounts were agreed to more enthusiastically than evolutionary accounts in the two younger age groups but not among the oldest group or the adults. Even more striking, however, rather than showing the muddled, roughly equivalent preference for spontaneous generation and creationist accounts, the youngest children significantly favored creationist accounts in the closed-ended questions over both evolution and spontaneous generation. Though roughly half of nonfundamentalist parents said they would teach a child evolution and none claimed they would teach spontaneous generation, the youngest children viewed evolution and spontaneous generation accounts as comparably bad as explanations for the origins of a first member of a species.

The fundamentalist children in Evans's study did show greater affinity for creationist accounts than the nonfundamentalist children. They preferred creationist accounts at all ages. Their parents almost exclusively advocated creationist accounts as well. The cultural environment and what children are taught matters to the sorts of beliefs they espouse about the origins of living things. Nevertheless, children up until about ten years old in both groups showed an affinity for creationist accounts over evolutionary accounts that cannot be accounted for by their parents' own position.

Perhaps most striking, this study suggests that evolutionary accounts as endorsed by parents or as included in school textbooks and science curricula do not penetrate children's thinking effectively until around age ten. Parents believe in evolution and say they would teach it, the textbooks include it, but the children just do not affirm it as might be expected. Why not?

Part of the answer may be a problem with the conceptual demands that evolutionary theory places on human minds. Evolution (whether by natural selection, supernatural guidance, or both) asks us to believe that one kind of living thing can change into another kind of living thing—that ancestors can have descendants that are a different species. The idea that a whale's great-great-great-great-grandparents and further back were cows violates our natural understanding of the relationship between parents and offspring—that they are the same species—and that living things have some kind of unseen, unchangeable essence that makes them what they are. A raccoon is a raccoon even if you surgically alter it to look like a skunk, and it will not have skunk babies.[18] Evolution asks us to believe something heavily counterintuitive, as do many scientific theories. Given its poor fit with the way human minds seem to naturally develop, it would be no surprise at all that people find evolution tough to embrace until fairly late in childhood, regardless of what those in authority might say.

Evolution has another problem when competing against creationist accounts for children's allegiances. Given the research discussed in Chapter 2 concerning young children's inclination to see the natural world as purposefully designed and their very early linking of order with intentional agency, creationist accounts are simply natural and intuitive. An evidently designed world demands a designer.

Children's tendency toward promiscuous teleology plus their understanding that intentional agents can order and design makes them prone to believe in a designer or creator. Their understanding of human limitations in terms of creative power eliminates people as a candidate for creator(s) of the natural world. If the idea of a nonhuman, creator God is proposed, children will find the idea very natural and intuitively plausible. It just makes sense to them that God created the heavens and the earth. Of course, we could imagine special cultural conditions in which children are drilled on evolution from an early age and repeatedly told that there is no god that accounts for any part of the natural world. In such extreme cases, children might learn to override their natural tendencies earlier, but barring such special cultural scaffolding, the natural default position seems to be for children to think a nonhuman someone or someones are the best explanation for the apparent purpose and order in nature all around.

Divine Superpower

Children readily embrace God as the creator of the natural world from mountains and lakes to trees and elephants, and this fact certainly testifies to children's readiness to accept beings with superhuman strength and power. Children's early interaction with physical objects impresses on them through direct experience that their own power to move objects and change their environment has limits. Doors that they want to open will not budge. Baby sister is just a little too heavy to comfortably lift and carry. Surely then, children can infer that someone who can create mountains must have amazing power generally. Right? Perhaps. Fortunately we do not have to take this position for granted without any independent evidence.

I have a dusty memory from childhood of a visit to my grandparents' home. I must have been around five years old at the time. A toy with which I had been playing managed to fly up and behind the enormous refrigerator in my grandparents' kitchen. Distressed at the loss of my toy, I asked my grandpa to please, please *pick up* the refrigerator so that I could get my toy back. I genuinely thought that my grandpa could simply grab the refrigerator and lift it. I had no doubt that he was strong enough to do so. After all, he had enormous biceps and could make them dance—it was one of his favorite tricks. My grandpa shook momentarily with his Santa-like laughter, hoisted himself from his recliner, went into the kitchen, and promptly retrieved the toy. (It was only on top of the refrigerator.)

What this memory suggests, provided I have placed it properly in my own development, is that around five years of age, I was beginning to understand the correspondences of musculature, size, and strength but had not yet learned the boundaries of human brute strength. I knew my much smaller grandma could not move the massive (from my perspective) refrigerator, but that my grandpa would have no difficulty with the feat. I was still learning human power limitations. For lessons about how even younger children understand human power, we turn again to Jean Piaget.

In his studies of children, Piaget concluded that they attribute superpower to all adults and because God is another adult human, to God as well. He wrote:

> There are many instances on record of children attributing extraordinary powers to their parents. A little girl asked her aunt to make it rain. M. Bovet quotes Hebbel's recollections of childhood. The child, that thought its parents all-powerful, was staggered to find them one day lamenting over the sight of their fruit trees ravaged by a storm. There was then a limit to his father's power![19]

As I have shared previously, for Piaget, the greatest evidence of children attributing superpower to humans was the case of humans' creating the sun, sky, mountains, and lakes:

> Not only is it evident that the omnipotence, with which the youngest of the children we have examined endow mankind in general, must be derived from the unlimited powers which they attribute to their parents, but furthermore we have often come across precise evidence in the shape of facts bearing directly on the point. We have fre-

> quently asked children if their fathers could have made the sun, the Salève, the lake, the earth, or the sky. They do not hesitate to agree.[20]

We now have evidence that children do not necessarily attribute the creation of the natural world to human beings but do have a strong tendency to see the natural world as purposefully designed. I am not aware, however, of any additional research that casts serious doubt on Piaget's claim that young children overattribute power to parents and to adult humans generally.

When my son was four years old, I made a short film with him illustrating the theory of mind research described in the next chapter. Just as I had done with other children previously, I began by asking him if he knew who God is. His response was telling: "Yup. He's the strong guy." For my son at that age, God represented strength and power. Though more systematic research on this topic would be helpful, at this point it is safe to say that children begin with a default assumption that other intentional agents are superpowerful, including their parents, grandparents, and gods. Maybe they are not all-powerful—that would be a sophisticated abstraction—but they are a safe bet to be able to do anything I (the child) can think needs doing and could possibly be done. Throughout the preschool years and even into the early school years, children learn the boundaries and limitations on different agents' power and strength. This first involves discovering that human power is limited and then specifying what good predictors of relative strength and power are, such as size and musculature. But learning just who can lift or move or do what takes a lot of time and mental effort. An all-powerful God is relatively easy to figure out. The superpower that children assume from the beginning needs no qualification or limitation. Can God lift or move it? The young child knows the answer even if she does not know what *it* is. God stands out from other agents in this regard. No wonder my son distinguished God as "the strong guy."

How Children Learn about Gods

I once knew a little girl who, at age two, astounded her parents by reporting that she saw angels in her room—up in the corner, to be precise. At age three, this same girl, in a fit of anger at her parents, stormed into her room, threw the door shut, and began muttering violently. I was there and witnessed the scene. When the parents opened the door and demanded that their daughter report what hostile slander she was spewing, she defiantly retorted, "I wasn't talking to you! I was talking to God!" Similarly, a mother of three-year-old Ben posted the following on Facebook: "Ben quote from yesterday: 'Water animals talk to God.' Also, 'Swimming dinosaurs talk to God.' We should all be so in touch. Of course those were followed up minutes later by a classic 3 year old line: 'I want to go to the store and buy toys.' "

I do not pretend that the angel-seeing girl is representative of children generally (and, incidentally, she has grown up to be a normal, well-adjusted person), or that little boys speculate about God's animal dialogue partners, but certainly the facility with which young children acquire and use god concepts is obvious. Similar to how they come to reason about other people, children from religious families easily form ideas about gods. They readily explain events as possible consequences of a god's activity. They make predictions and suppositions about gods' thoughts, opinions, and wishes. They apply ideas about gods in novel and sometimes personal ways. From where does this religious fluency come?

The easy answer (perhaps too easy) is that they are taught it: children believe because their parents (and other trusted adults) act as if they believe, and talk as if they believe. Until given strong reason to believe otherwise, this testimony is powerful. We might call this the *indoctrination hypothesis.*

But such an account of how ideas are passed on to the next generation would be overly simplistic if we did not consider the character of the beliefs being communicated and how well they are accommodated by children's minds. Anyone who has taught children, as a parent or a teacher, knows full well that children cannot be easily taught just anything. Some ideas seem to stick better than others (as Evans's research on creationist versus evolutionist beliefs illustrates). Further, though some parents may carefully indoctrinate their children and threaten grave repercussions for disbelief, most religious belief in childhood seems to be more simply absorbed. The question to be answered, then, is how and why many religious ideas are so easy for children to adopt.

One answer to this question has been at the theoretical center of the scientific study of religion and especially in the psychology of religion for over a century.[21] This answer might be called the *anthropomorphism hypothesis.* As *anthropomorphism* means making something in the image of people, this hypothesis is the opposite of the biblical idea of people being made in God's image: God was, is, and is being made in the image of people. By this view, children learn about people—what they think, how they act, what they like—and then analogically reason about gods. Learning god concepts is a simple matter of learning about people and then applying that learning to gods (though how simple this process is remains unclear). No wonder children find thinking about gods, ghosts, spirits, and devils so easy.

The anthropomorphism hypothesis further maintains that through the course of development, gods look less and less like a human because children have more sophisticated reasoning abilities to draw on to make sense of them. For Christian, Muslim, or Jewish children, God begins as a big person living in the sky and then either gradually or radically becomes an all-present, formless, unchanging, nontemporal, all-knowing, and all-powerful being. Crude anthropomorphism gives way to God as an abstract being with unusual properties.

In contrast to the anthropomorphism hypothesis, I argue that children do not have to reason about gods as they reason about humans. In fact, children's minds actually facilitate the acquisition and use of many features of God concepts of the Abrahamic monotheisms (Judaism, Christianity, and Islam) and perhaps those in other traditions as well.[22] I have called this proposal the preparedness hypothesis.[23] Children may easily form ideas of God because their mental mechanisms have two properties that favor learning about God. First, this equipment easily entertains nonhuman agents. Second, it appears to presume superhuman properties until it discovers otherwise. Superagents fall close to natural default settings.

Recall that in Chapter 1, I shared evidence that children's mental tools for detecting agents in the environment can identify things as agents that do not remotely resemble human beings. Colored circles will do the trick. Similarly, when it comes to reasoning about the minds of these agents, children's relevant conceptual systems do not require humanness. Rather than being solely dedicated to helping understand humans, they operate generally on any and all intentional agents. Hence, children are initially capable of understanding lots of beings—from God to ghosts to gorillas—as well as they understand humans, and supernatural properties do not impose undue conceptual burdens.

A second feature of children's minds that may favor the acquisition of ideas about some gods is that their minds assume that many superhuman properties are the norm, simply by default. For example, when the hypersensitive agency detection device (see Chapter 1) identifies something as an intentional agent, a three-year-old automatically assumes that the agent has the superhuman property of full access to information or infallible beliefs (at least within certain limits discussed in the next two chapters). Developmental psychologists continue to find evidence that the godly properties of superknowledge, superperception, creative power, and immortality are quite intuitive, at least for young children. Concepts of God are easily accommodated because they play on many of these default assumptions rather than violate them. In Chapters 4 and 5, I detail some of this evidence with regard to children's understanding of God's superhuman mental and perceptual properties, immortality, and goodness.

FOUR

The Mind of God

IN *The Hitchhiker's Guide to the Galaxy,* Douglas Adams reveals that planets are designed and built by teams of space aliens that look and act quite similar to us humans. They just happen to have the skills and technology to build entire worlds:

> "Are you trying to tell me," said Arthur slowly and with control, "that you originally . . . *made* the Earth?"
>
> "Oh yes," said Slartibartfast. "Did you ever go to a place . . . I think it was called Norway?"
>
> "No," said Arthur, "no, I didn't."
>
> "Pity," said Slartibartfast, "that was one of mine. Won an award, you know. Loverly crinkly edges."[1]

Could this be how children view the agency behind the natural world? We have seen that they know people—human beings—did not make the world, but could it be that the gods that create are not all that different from humans with the ability to create natural things but susceptible to other human-like limitations?

For many adults, gods typically possess other characteristics besides being responsible for some (or all) of the natural world and natural occurrences. Adults' gods sometimes have features such as unearthly knowledge; seeing, hearing, and attending to much more than people can manage; and being immortal.

In this chapter I consider children's developing understanding of mental properties and how this development might facilitate or even encourage children to learn about and understand gods with superknowledge. In some respects, getting a handle on other human minds and animals' minds may present more challenges than a basic understanding of God's supermind.

God as a Superknowing Being

One prominent area of research in child development over the past two decades has been an area called theory of mind. This theory concerns how and when people understand others' minds, including their thoughts, perceptions, wants, and feelings. One reason for which this research area has been so prominent is that theory of mind is critically important to normal social functioning.

Imagine for a moment if you did not understand that other people have mental states such as thoughts or emotions. Other people would be no more than complex machines that often exhibit inexplicable behaviors. Their feelings, recollections, hopes for the future, opinions, and yearnings would escape us. Simple human social exchanges would be baffling.

Consider the following scene: John wakes up early in the morning and sweeps into the kitchen. He orients his head and eyes toward the coffeepot in the corner, and a slight smile dances on his lips. John makes coffee and then goes around the corner in the next room to eat breakfast, not drinking any coffee himself. John's younger sister, Jane, comes stumbling into the room and begins rummaging for breakfast foods. Soon John and Jane's mother enters sniffing the air. Mother orients her head and eyes toward the coffeepot and a broad smile spreads across her face. "Thank you, Jane, for making me coffee," Mother says. Without looking up from her food scrounging and with a flat monotone, Jane utters, "You're welcome." John's face changes expression, his eyes squinting and his lips tightening. He rushes out of the room with a hunched posture.

This scene presents a number of questions for the observer. Why did John smile at the coffeepot? Why did John make coffee that he did not drink? Why did Mother thank Jane for making the coffee? Why did Jane say, "You're welcome"? Why did John leave the room suddenly with a change in his facial and bodily posture?

To those with normal, mature theory of mind capabilities, such a scene is easily understood as involving a host of different mental and emotional states that somehow make sense of what happens. John (perhaps) thought he would make coffee to please his mother and smiled at the thought of pleasing her or her future reaction. Seeing Jane, Mother assumed (incorrectly) that Jane made the coffee. Pleased, because she likes coffee in the morning, Mother thanks Jane. Jane receives the thanks either because she wants credit for doing something she did not do or because she groggily was not really paying attention and answered reflexively. John's facial and bodily postures change because he felt negative emotions because Jane received credit for something he did or because he perceived Jane as deliberately giving the impression that she had made the coffee.

Most adult humans take for granted our ability to rapidly understand the feelings, wants, thoughts, and perspectives of others and how they motivate action, not realizing what a tremendous accomplishment such abilities are. This robust theory of mind reasoning may be exclusively the skill of human beings, and not even all humans. People suffering from some disorders, such as autism, may not have these abilities readily at their beck and call. And even among normally developed adult humans, the rapidity and fluency with which some reason about the thoughts and feelings of others varies considerably. Some are socially adept empathizers and need very little information to draw rich inferences about others' thoughts, feelings, desires, personalities, and relationships. For these social geniuses, the following four-sentence soap opera–like exchange is enough to spin a whole story about life, death, and love:

"Is he?"
"Yes. I'm sorry."
". . . I guess I'll have to move on."
"Can I buy you some coffee?"

And then there are others who just do not get it even with far more abundant information. Although they may be brilliant in other domains such as philosophy or the sciences they are socially disabled. (Universities seem uncommonly blessed with such folk.) Similarly, children before the age of four may be bereft of these abilities—or at least lacking great facility with them. Theory of mind reasoning is a critical developmental achievement.

Many of us have had the experience when playing hide-and-seek with a two-year-old that the child will hide in plain sight, covering only her face. I have seen children of this age hide by simply standing in the corner of a room, their faces to the walls. Such a strategy evinces a lack of awareness of others' visual perspectives and resulting knowledge about where the child is hiding.

Have you ever been on the phone with a three-year-old child and the child starts talking as if you are there in the room with him? He tells you, "This is my new shirt, and that is my kitty, and he's my friend Bobby," assuming that you can see what he is talking about. Again, this amusing tendency is evidence that children of this age find it difficult to keep track of what others see and know. They assume that if they see it or know it, you must too.

The cumulative evidence from decades of research on how children understand others' mental states (including perspectives, beliefs, and desires) demonstrates that two- and three-year-old children do not fully understand the nature of others' beliefs and perspectives. Though new experimental evidence within the past few years has raised the possibility that children have a rudimentary theory of mind in their second year of life, the bulk of evidence shows that before age four or five, children might be considered self-centered realists when it comes to other people's mental states.[2] They appear to consciously think, *Whatever I know is the case, is also what others know is the case.* If I, the three-year-old, know that there are worms in the silverware drawer *(I wonder how those got there?)*, my mother knows it too.[3] If I, the three-year-old, can see the bird out the window from where I sit, you, on the opposite side of the room away from the window, can see the bird too.

If you get the chance, try the following game with a young three-year-old. Get a familiar container such as a saltine cracker box. Empty it of crackers, and place some surprising objects in it, such as pencils or rocks. Then close the box. Show the three-year-old the closed box (be careful that it does not rattle or make a strange noise that gives away that you have tampered with it). Ask her what she thinks is inside the box. Odds are, if she is familiar with the box, she will say, "Crackers." Then open the box and let her see the rocks inside the box. She will be surprised and look at you with a mixture of disappointment and astonishment. (You must be a *very* strange person.) Close the box up again, and get the child to agree that she can no longer see what is inside the box. "Okay, I've closed the box so that no one can see inside it. Can you see inside this closed box?" Then ask, "Do you remember what is inside of this closed box?"

After she answers, "Rocks," again, the fun may begin. Pick someone the child knows who has not been in the room (for my illustration, I call this person Mary). Then ask, "If your friend Mary came over to play and she saw this closed cracker box, what would she think is inside of it?" It is likely that the three-year-old will answer, "Rocks." If you then ask her to explain why Mary would think there are rocks in a cracker box, the child will probably simply point out that there *are* rocks in the cracker box. That is reason enough for Mary to think there are rocks in the cracker box. In other words, the fact that there are rocks in the box makes it hard for her to say that Mary does not know what is in the box.

Even more astonishing, ask the young three-year-old what she thought was in the box when you first showed it to her, before you showed her what was inside it. She is likely to answer, "Rocks." (Try not to laugh.) Ask then if she ever thought there were crackers in the *cracker* box. Chances are, she'll say, "No." Before jumping to the conclusion that this cute and delightful child is a horrible liar, realize that children of this age do not have terribly strong access to their own mental states. That they currently believe there are rocks in this box makes it difficult for them to understand that they previously thought otherwise—they are realists: whatever is the case is what other people think is the case. Most young three-year-olds act this way. (Try it out sometime—with a parent's permission, of course.) By age five or six, most children understand that others looking at the box for the first time would be fooled by the appearance and falsely believe it to contain crackers. These older children also will know that they themselves had been fooled.

The series of questions just described is one type of false-belief task that developmental psychologists have used to explore children's understanding of others' beliefs. It is called a *false* belief task because it checks to see whether children understand that beliefs can be false. In the example, the child did not understand that Mary had a false belief about the contents of the cracker box. It is not until they can tell that an idea could be mistaken but still be believed that we can be confident that children really understand what beliefs are. Before that point, "I believe . . . ," could simply mean, "It is the case that . . ." I return to the cracker box task because my collaborators and I used it to look at children's understanding of God's beliefs.

God's Knowledge

Across all theory of mind experiments, one theme arises again and again. Three-year-olds and many four-year-olds, depending on the task, find it difficult to understand the perspectives and beliefs of other people. As with the cracker box, they think that people see the world the way they know it to be. Others always have accurate knowledge of the world (at least as the child believes the world to be).

It occurred to me and my former students Rebekah Richert and Amanda Driesenga that this understanding of others' beliefs as always accurate sounds a lot like how Christians, Jews, and Muslims think about God. God's beliefs are always accurate too. Based on the conventional wisdom handed down from Piaget and his followers that children cannot distinguish between God and human beings until they are about eight,[4] children would go from thinking about God as having accurate beliefs just like a human (age three) to thinking about God as having mistaken beliefs just like other people (age five) and back to understanding God as having accurate beliefs when they can understand God is not just a special human (around age eight). This is the kind of irregular developmental pattern that gets developmental psychologists excited, so we thought we would check it out. The questions before us were: *Do children really treat God's knowledge as identical with humans' knowledge? Is it even possible for children younger than age eight or nine to do otherwise?*

To address these questions, we conducted the cracker box task with a number of three- through six-year-olds from Protestant families (we needed kids who knew what "God" meant). After presenting them with the closed box, asking what they thought was inside the box, showing them the rocks inside, and then closing the box back up, we asked, "If your mom [or dad] saw this closed box for the first time, what would she [he] think was inside it?" We also asked the same question regarding several other potential viewers such as various animals and God. "If God saw this closed box for the first time, what would God think was inside it?"

When answering about their mother (or father) or any of the animals, children showed the same pattern as we have become accustomed to in false-belief tasks: around 80 percent of three-year-olds said mom would think there were rocks in the box. By five years old, around 80 percent of kids answered that their mom would think there were crackers in the box—a false belief. If the Piagetian idea is right that "God" can only mean a particular human being for children of this age, then we would expect the same pattern of answers when children considered God. That is not what we found. Rather, children in all age groups were equally likely to say that God would know there were rocks in the box—roughly 80 percent or more of the children in all age groups. So here is some evidence that children do not have to reason about God as just another human. Already at age five, children treated God as significantly different from people (and animals) in the false-belief task. Children were not simply aping that "God knows everything" but apparently used their understanding of God's superknowledge to solve a strange new problem.

Younger children did treat God as similar to human beings—not by making God trickable like a human, but by treating humans (and animals) as untrickable like God. In this population of children, we saw that from three years old, children were theologically accurate (God would know there were rocks in the box). The majority was not accurate regarding humans until five years old.

So Piaget's idea that children cannot conceive of God as different from a human being until around age eight or nine is mistaken. Nevertheless, Piaget's reasoning might not have been all that far off the mark. One reason he saw for the alleged confusion between children's understanding of God and humans was that they attributed Godlike properties to adults and then spent childhood learning humanly limitations. In this respect, Piaget may have been right. From this experiment and others like it, it appears that the three-year-olds treat God and humans similarly, not because God is human-like but because people are, in certain respects, Godlike in terms of knowledge.[5]

Another set of experiments makes the same point in a different way. In these experiments I investigated what might be called *background knowledge*—previous knowledge required to understand or interpret an otherwise unclear visual display or activity.[6] A superknowing God should have any and all needed background knowledge. An easy way to understand the motivation for these experiments is to consider the perspective of the family dog or cat.

When I was a child, my parents took my brother and me camping in the Sierra Nevadas. They also brought along our faithful family dog, Dobby—a golden retriever and clumber spaniel mix. As my more temperature-sensitive parents watched from the shore, my brother and I swam and played in the crystal-blue mountain lake. Before long, our squeals of delight made our increasingly anxious dog jump into the chill water and attempt to save us. She was certain we were in grave danger and required her rescue. Now perhaps I am attributing too much mental activity to this fine animal companion, but I hope the illustration is clear. The dog misunderstood what she was perceiving, not because she could not see or hear what was necessary (as in the cracker box task), but because she did not possess the needed background knowledge to make sense of what she did see and hear. You could put a newspaper in front of an animal and it would not be able to read. Likewise, it could see you with a book in front of you and not know you were reading. You can probably imagine a number of other cases where your beloved cat or dog could see what you were doing but could not possibly understand it.

Animals are not the only ones that sometimes lack the background knowledge for understanding what they see. Children and adults have this problem all the time. If you have ever heard someone speaking an unfamiliar language or tried to read something in a foreign language, you have confronted the same problem: you did not have the background knowledge to make sense of what you heard or saw. As adults we know that experiences and general capacities color and inform how we understand present experiences.

The questions these observations raise for the developmental psychologist interested in theory of mind are: *When can children understand that not all people have the same background knowledge? Specifically, do they distinguish among people, animals, and God as more or less likely to have background knowledge relevant for understanding a display or event?*

Roxanne Newman, Rebekah Richert, and I began exploring these issues experimentally through a series of three tasks requiring children to understand the nature of background knowledge for making sense of displays.[7] The first task was a variation of the previously used droodle task.[8] In this task, the experimenter shows the child a drawing partially covered by card stock. Through a window cut out of a larger sheet of card stock, the child sees a red circle flanked by two blue rectangles. The experimenter asks the child what the covered picture depicts. The child answers (for example, "A big red circle and two little blue rectangles"). Then the experimenter uncovers the picture, bringing it into full view: a drawing of two blue elephants holding a red ball between their trunks. The experimenter checks to see that the child knows what the whole drawing depicts now that she has full visual access and then recovers the picture with the card stock so that it appears as it did initially (showing only a red circle and two little blue rectangles). The experimenter then asks the child to consider what someone seeing only the covered drawing would think it depicted.

The second task was a secret code. We created three strange-looking symbols to be part of our code. The experimenter said that she had invented a secret code that no one else knew about but that she would share it with the child. She then placed the three sheets of paper, one with each symbol, in front of the child and explained what each meant (for instance, ball, bicycle, and so on). She then checked to see whether the child could say what each symbol stood for. Once the child learned which symbol went with each word (it did not typically take long; they got it right on the first try), the experimenter asked questions about what other people would know if they saw the code for the first time.

The third task was based on the same logic. Have you ever had the experience of entering a room where a group of people were excitedly and cheerfully playing some kind of game that you could not understand at all? I have. I watched and listened for a few minutes and deduced that a game was being played but had no idea what motivated people to do what they did and say what they said. I lacked the background knowledge to make sense of the rules. As frustrating as it can be, knowing that a game is being played is a conceptual accomplishment. A typical house cat or even an infant probably would not understand even that much. In this third background knowledge task, Newman created a new board game with a homemade board, pieces, and utterly opaque rules—unless they were explained to you. She then started playing the game in front of a three- to six-year-old child; after a while she stopped playing and asked, "Do you know what I am doing?" After establishing that the child did not know what she was doing or what the rules of the game were, Newman explained that it was a new game she had invented and that no one else knew it. She then explained what the rules were and invited the child to play along. Then came the questions about another person, a dog, and God.

Across these three tasks—the droodle, the secret code, and the novel game—three-year-olds tended to think that their mother, God, and anyone else would know what the covered picture showed, what the secret symbols meant, and what the game was. Similar to the cracker box task, five- and six-year-olds answered consistent with the understanding that no one except God would know what the experimenter was doing before the explanation. Also similar to the cracker box task, younger children overattributed knowledge to the person and dog, treating them as godlike. Three-year-olds assumed others would understand the game; the majority of children at all ages thought God would understand the game. The Piagetian prediction was wrong. Children younger than eight years old can attribute true beliefs to God and false beliefs to humans.[9]

In Chapter 1, I mentioned Bradley Wigger's research on children's imaginary friends.[10] He asked three- to five-year-olds with imaginary or invisible friends to complete the droodle task, the secret code task, and the cracker box task. He asked each child about a visible friend of the child, the child's invisible friend, a dog, and God. Results were similar to those from the studies described. The youngest children treated all the agents as superknowing, but the older children treated only God as superknowing. Interestingly, the children began attributing false beliefs or ignorance to the dog and the visible friend before the invisible friend, and even the oldest children regarded their invisible friends as more likely to know things than their visible friends or a dog. Invisible friends were more godlike than the visible friend and dog. Without indoctrination or a religious community supporting them, these children had invented their own invisible, superhuman agents: gods.

God's Memory

From courtroom testimony to deciding whether the show time was 7:00 or 7:30, we often have to decide whether to trust another person's memory. Considerations of who simply has a better memory than another may come into play. My wife, for instance, has a fine memory for many things, but numbers is not one of them, whether the numbers are times, dates, quantities, or something else. Contrary to popular stereotypes, it is she, not I, who cannot seem to remember the date of our wedding anniversary.

Watching animals, too, we might observe that different animals have different abilities to remember things. Squirrels forget where they hid their acorns, but birds seem to know exactly where they are. Elephants, we are told, never forget. Infallible elephant memory may be an exaggeration, but do gods forget?

My collaborator Emily Burdett designed an experiment to examine children's attribution of remembering and forgetting to others, including God.[11] The task used the common memory card game in which children are shown a number of cards with drawings on one side and nothing descript on the other, and asked to remember where the animals (or whatever is drawn) are once the cards are turned over. Burdett presented memory cards, to three- to seven-year-old Israeli children from practicing Jewish families, adding to the number of cards until children could no longer remember where a target animal (say, a cow) was located when the cards were turned over. After the child admitted not remembering where the animal was, Burdett asked whether various others, including Mom and God, would remember in the same situation. In this case, children of all ages, even the youngest, thought God was likely to remember when they had forgotten and was more often regarded as remembering than their mother. But as in previous studies, children tended to give others' knowledge the benefit of the doubt. When they were ignorant themselves, they still often thought others (animal, human, or divine) would know.

Is This Just Indoctrination?

These studies teach us three things. First, God is not just another human for children younger than eight or nine, as Piaget thought. Rather, by age five, clear distinctions are possible. Second, in these scenarios, three- and four-year-old children more accurately predict God's knowledge (according to their parents' Christian theology) than other humans' knowledge, tending to overattribute knowledge. Third, by age five, their accuracy when considering humans catches up to that when considering God.

If you have been indoctrinated by the indoctrination hypothesis or if you are just a healthily suspicious person regarding overenthusiastic conclusions, you might have picked up on the fact that the children in the studies reported in the previous chapter were from religious families. Perhaps something peculiar about children raised in such families enables them to "get God right" at so early an age (and treat their parents as godlike). For instance, maybe they have had the notion of an omniscient God so drilled into their heads that they automatically and appropriately apply this theological teaching to the context of these strange little games concerning what God knows. Perhaps.

I concede that these children might not be representative of children generally in terms of both their educational and religious upbringing. Perhaps these children only provide evidence that kids younger than eight need not anthropomorphize. That is, four- to six-year-olds are capable under some cultural conditions of treating God's perception and knowledge as distinct from that of humans, but these may be fairly special conditions.

If the indoctrination hypothesis is correct, children have no special receptivity to the idea of a god with superattributes and just learn whatever they are taught with comparable facility. So if children are taught religions in which some gods are just as fallible as people, whereas other gods are supersmart, they should learn about both types of gods at the same pace. That is, if children have no natural biases toward supersmart gods, dumb gods should be learned as early or earlier. Such a polytheistic population with both smart and not-so-smart gods has been investigated, and the indoctrination hypothesis seems to have a problem.

In spite of the impression with which you might have come away from grade school history classes, the Maya are not extinct but do live today in southern Mexico and in Guatemala. Many Maya have a syncretistic religion—their beliefs are a blend of traditional Maya beliefs and the later-arriving Christian ones, primarily of the Roman Catholic variety. Consequently, they have many different gods in which they believe. They believe in Diyoos (the Christian God), the sun god, forest spirits (called Masters of the Forest), and a number of other supernatural beings including the Chiichi'—something comparable to the bogeyman. Diyoos is all-knowing. The sun god, with its lofty vantage point, knows about whatever goes on under the sun, so to speak. Consequently, this is also a very knowledgeable deity. Wherever light is, the sun god's knowledge reaches. As their literal name indicates, the Masters of the Forest are spirits who know all there is to know about the forest, including its plant and animal inhabitants and what takes place in the forest. If you are injured in the forest or are hunting in the forest, it is to the Masters of the Forest that you should make your petitions. The Chiichi' are minor spirits. Like a bogeyman or other minor spooks that adults might use to frighten children into right behavior, the Chiichi' are fallible in their powers and knowledge. They can be disgusting or annoying but can be tricked and fooled. This diversity of gods with their various properties afforded an excellent opportunity to try to see if an importantly different group of children could make similar distinctions between a god and humans and to see if these children are able to distinguish among various types of gods as adult Maya are. Diyoos is all-knowing, but the Masters of the Forest's expertise lies in the forest only. Can children make these finely grained distinctions?

My collaborator, anthropologist and psychologist Nicola Knight, adapted a version of the cracker box task to use with Maya children.[12] As cracker boxes are not readily familiar to these children, for the familiar container he selected a hollow gourd commonly used to house tortillas. Finding some rocks in such a gourd might not be nearly so unexpected or surprising as finding them in a North American cracker box, so Knight used a truly surprising object: boxer shorts underwear.

With the help of a translator using the children's first language of Yukatec, Knight interviewed children ranging in age from four to seven years old about the tortilla gourd and a number of agents: Diyoos, the sun god, the Masters of the Forest, the Chiichi', a doll representing a human, and several familiar animals represented by puppets (bee, dog, peccary, and jaguar). The setup was similar to that in the cracker box task. The experimenter showed children the gourd with its opening covered by a piece of cardboard, asked what they thought was inside it, and then showed them the surprising contents of boxer shorts. The gourd's opening was recovered, and then the experimenter asked what the human puppet, the animals, and the various deities would think is inside the container. As in previous studies, the youngest children treated Diyoos and the human similarly. A majority answered that both would think that there were boxer shorts in the gourd—the same gourd that the child believed would contain tortillas and surprisingly contained shorts. By age seven, the majority thought that Diyoos would know that the gourd contained shorts but the human would mistakenly believe it contained tortillas. As with the American version, while answers significantly changed with age regarding what the human would know, children showed no age differences when answering about Diyoos. Old or young, children were equally likely to answer that Diyoos would know boxer shorts were in the gourd.

The first question, whether the ability to distinguish between God and humans before age eight was peculiar to the

American sample, has been answered. Both the Yukatec Maya and the Americans distinguished between humans and God as soon as they could successfully predict human beliefs. Before "passing" the false-belief task, they treated God and humans similarly, attributing to both of them true beliefs.

What then about the other gods? Surprisingly, children showed some sensitivity to the shades of difference in potential knowledge of the various deities.[13] A majority of children thought Diyoos would know the contents of the gourd, but only about half the time they answered that the sun god and the Masters of the Forest would get it right and think shorts were in the gourd. When it came to the dumber Chiichi', children's answers were similar to when asked about a human or the animals. They most often thought the Chiichi' would believe the gourd contained tortillas. This experiment provided evidence that children not only can distinguish gods from humans before age eight, but also that they can begin to distinguish among various gods at these early ages. When they could understand the nature of beliefs as potentially false, Maya children showed they could likewise discriminate among the shades of potential knowledge of various deities. With all agents, however, the younger children tended to attribute superknowledge. Even if they were indoctrinated that Diyoos knows everything, surely they were not taught this about the Chiichi', and yet for the younger children, the Chiichi' (as well as people and peccaries) would know what was in the gourd as well as God would. These results fit with the born believers thesis but are difficult to account for with the indoctrination hypothesis.

What Does God Know?

Three-year-olds attribute superknowledge to gods, but I do not mean to imply that these children understand God and others as omniscient or all-knowing. Indeed, we have some reason to believe adults do not necessarily view God as all-knowing, strictly speaking. Conceiving what it would mean to be all-knowing is conceptually difficult (at least it is for me). I find it difficult to fully grasp what it would mean to know that which I do not even know is a thing that can be known. (See what I mean?) The results of the cracker box experiment, tortilla gourd experiment, and the others described leave open a number of possible interpretations. What do these youngest children think gods' knowledge is like? Here are three possibilities:

1. *Omniscience.* Children assume that other beings know absolutely everything. That is, God, their parents, and even elephants are all-knowing or omniscient.
2. *Child-centered knowledge.* Children assume that others know everything that they themselves know but nothing that they themselves do not know—sort of a generous egocentrism. A subtle variant of this option is that others' have perfectly accurate beliefs only about what children themselves care about but not necessarily what they currently know.
3. *Full access.* Three-year-olds and younger children assume that others do not know absolutely every-

thing but only what children recognize as knowable or worth knowing.

More experiments would be helpful in distinguishing which of these three possibilities most accurately captures what three-year-old children think about others' beliefs, but I think it is safe to rule out the first option. That children presume that others know absolutely everything—are omniscient in a strict sense—is unlikely. After all, knowing that others know absolutely everything may require a fairly good grasp of what it might mean to know everything—particularly things that we ourselves do not know or even know are things that could be known. Some experiments conducted with Greek and Spanish children challenge full-blown omniscience (option 1).

Nikos Makris and Dimitris Pnevmatikos presented Greek children from Orthodox Christian families with a box containing an unknown object and asked whether God and a puppet (Titi) would know what was in the box.[14] The result? A majority of three- and four-year-old children answered that God would not know what was in the box, a similar response rate as for Titi. By age five, children clearly distinguished between God's knowledge and Titi's, with a majority of children thinking that Titi would not know what was in the box but a majority believing God would know. As before, children distinguished God's mind from a human's years before what might be predicted based on Piaget's perspectives, but in this case, the youngest children did tend to inaccurately predict God's knowledge. Such a finding could be particular to the population in question or the task used (more research is needed to be confident), but it also could indicate that children do not think of God as all-knowing, but only superknowing in some respect (option 2 or 3 above). Perhaps young children in a state of ignorance find it difficult to see the unknown as knowable. Because they lack their own knowledge to drawn on, they may find it more difficult to attribute knowledge to others.

Marta Giménez-Dasí and colleagues conducted a similar experiment relevant to whether children understand God (and others) as fully all-knowing versus only superknowing.[15] Giménez-Dasí's team presented Spanish children with a variation of the cracker box task in which the familiar container (in this case, a Smarties candy container) was initially wrapped like a present, concealing any information about its possible contents. The experimenter asked half of the three- to five-year-old children whether a friend would know for sure what was in the box when it was still wrapped or would have to guess. They asked the other half of the children the same question with regard to God. Would God know for sure what was in the wrapped box before the lid was removed? In Giménez-Dasí's study, three-year-old Spanish children tended to think God would know what was in the box, whereas a friend would not. Over three-quarters of their answers to the two questions affirmed God would know what was in the box, but roughly two-thirds of the time they said their friend would not know. These general patterns persisted with five-year-olds too. Perhaps, then, even three-year-olds are capable of recognizing that a friend's ignorance does not apply to God and are able to distinguish human minds from divine ones.

Emily Burdett and I revisited these issues with a series of tasks she administered to Jewish Israeli children, including another unmarked box task with unknown contents. Even when ignorant themselves about the location of a particular card (as described previously) or of the contents of a box, the youngest children (three- to four-year-olds) most often assumed that God would know. That is, even when they themselves did not know what was in the box, they could still successfully attribute knowledge to God. These findings cast doubt on option 2 and support option 3 that children have a tendency to treat gods as full-access superknowers.[16]

Though many questions remain to be answered, we may tentatively conclude that three-year-olds (1) are capable of distinguishing between gods and people in terms of what they might know and not know, (2) have a tendency to overattribute knowledge and accuracy of beliefs to others in many cases, but (3) it is unlikely they attribute full-blown omniscience to gods or humans since they may regard them as ignorant of reality in some cases (particularly when the child is ignorant too). It seems that instead of omniscience, three-year-old children grant superknowledge to all agents: God, their parents, and others, they think, know all that there is to know (full-access knowledge). Children begin with a superknowledge assumption that in many instances is more godlike than human-like.

These findings suggest that it might not be difficult at all for kids to learn about and believe in a superknowing God. In fact, it may be harder for kids to learn about and believe in a limited, often ignorant parent. The good news for parents is that they learn God is smart sooner than they learn we are dumb.

FIVE

The Nature of God

THOUGH PEOPLE in the English-speaking world commonly think of gods as having special smarts and access to what people do in secret (through seeing all, hearing all, and mind reading), not all gods come in this variety. Some gods either have their knowledge limited to specific places and activities or they just are not very smart. The forest spirits of the Yukatec Maya have impressive knowledge about forest life and what happens in the forest but little regard for what goes on in the village. Similarly, in his ethnography of the Baining people of Papua New Guinea, Harvey Whitehouse notes that though the ancestors are regarded as knowing people's thoughts and activities, the forest spirits, or *sega*, are regarded as fallible and not having special access to people's minds.[1] Within Christianity we see these distinctions between supernatural beings. As a child, I was taught that God knows everything: facts about the world, the future, our thoughts, and our behaviors. Satan certainly was shrewd, but could not read minds and could make mistakes. Similarly, in traditional Jewish practice, there is a monthly ritual blessing of the new moon. During the month of Tishrei, however, the observance is delayed until after Yom Kippur observances. A common folk explanation for this deviation in regular practice is so that Satan does not know the exact date of Yom Kippur. Apparently Satan is a little dim to be so easily duped year after year.[2] Some deities and supernatural beings are dumber still. In many traditional Asian and African cultures, parents give babies misleading names and even insult them so that malevolent spirits do not recognize and attack them.

The tendency documented in Chapter 4 that young children have toward assuming all others have superknowledge, then, assists them only in acquiring certain types of god concepts with particular properties. Not all of the divine attributes commonly ascribed to God in Christianity, Islam, or Judaism receive special developmental treatment. We have no evidence that being three persons in one (the Christian notion of Trinity), for instance, makes clear sense to children or many adults, for that matter.[3] Similarly, a god having no location in space or time, as suggested by some Muslim, Jewish, and Christian theologians, appears to strain the minds of adults and children alike—a point I return to in the next chapter.

Just which divine properties come naturally to children remains an open question. In this chapter I consider several additional attributes. Primarily, I consider whether children may easily grasp God as superperceiving, immortal, and morally good by virtue of how children's minds develop. Perhaps the god in which children are born believers is supersmart and somehow responsible for designing at least some part of the natural world, but is otherwise like an ordinary person—more like Superman than God.

God's Perception

Children's understanding of perception is often treated as part of theory of mind research because what we know about the world around us is inextricably tied to what we learn through our senses. Adults know this, but when do children come to know that seeing typically leads to believing? Or that there are some things we humans cannot know without seeing, hearing, or otherwise perceiving?

As we know from Chapter 4, two- and three-year-olds have difficulty with a number of problems related to the visual perspectives of others. They appear to think you see things the way they do. So if a picture book with an illustration of a house faces the child and you are sitting opposite him, the three-year-old assumes you see it right side up too instead of upside down. When asked to select the picture of what the person across the table sees from where that person sits, children at this age tend to select the right-side-up drawing (or upside down, if that is what the child sees). Two-year-olds and many three-year-olds have trouble even telling when another person can see what they see.

A study by John Flavell and colleagues illustrates how dramatic this deficit is.[4] Experimenters sat three-year-olds at a table with a cardboard screen. Each child was presented with two cups, a white one and a blue one, and asked which one was white and which one was not white. The white cup was then taken away, leaving the blue cup, and the experimenter assured the child that the blue cup was not white. (Duh!) Another experimenter (Ellie) then entered the room and sat across the table from the child, on the other side of the cardboard so that she could not see the cup. The experimenter then asked the newcomer, "Ellie, we have a cup over here. Do you think we have a white cup over here?" Ellie then replied, "I can't see the cup. Hmm. I think you have a white cup over there. I think you have a cup that is white." The child was then asked if Ellie thought the cup was white, indicating an understanding that Ellie had a false belief. Note that in order to get this question correct, the children merely had to accurately repeat what Ellie had just said: that she thinks the cup is white. Three-year-old children answered this question incorrectly two-thirds of the time, saying that the cup is blue. The fact that the cup really was blue made it hard for them to imagine (and say) that anyone else could think otherwise.

These perspective-taking findings mirror results from the cracker box task described in Chapter 4. Three-year-old children have a tendency to (erroneously) extend their visual perspective to other humans. Might they likewise readily assume that gods can see, hear, or smell everything they themselves see, hear, or smell?

My mother shared with me that during her fundamentalist upbringing, she found creepy the emphasis on "God is always watching." She confessed that after a particularly rigorous encounter with, "You'd better behave 'cause God is always watching," she felt very uncomfortable using the toilet because of peeping-God.

Certainly being able to see and hear more than what other humans can see and hear is commonly attributed to gods around the world—even to gods that are not all-knowing. Because of being invisible or having special perceptual abilities, gods frequently see and hear more than even the snooping next-door neighbor. Indeed, this access to information, particularly information about what we say and do, enables gods in many cultures to be enforcers of human morality or to punish offenses against the gods and their territory. My collaborators and I wondered, then, whether young children can reason about God as having different perceptual abilities.

Under cover of darkness, lots of bad things happen unseen. At least that is what books and movies of this title lead us to believe. But are they really unseen—or just unseen by human eyes? Developmental psychologist Rebekah Richert and I devised an experiment in which we asked three- to eight-year-old children about seeing in the dark.[5] The experimenter showed each child a box with a slit in the top of it to peer through. The child was invited to look through the slit and was then asked, "What do you see inside the box?" After the child reported seeing nothing, the flashlight mounted on a hole in the side of the box was turned on, illuminating a wooden block inside. The light was turned off, and the child was invited to look again. Then the experimenter explained that cats can see in the dark because of their special eyes and asked the child what a human puppet, a cat, a monkey, and God saw in the darkened box. While most three-year-olds reported that the human puppet could see the block in the darkened box (which had been invisible to themselves), only a minority of five-year-olds did so. In contrast, a large majority of the children of all ages answered that God and the cat would see the block—up to 90 percent in some cases—a tendency persisting with the older children. Similar results were later found with children from Orthodox Christian families in Greece.[6]

Visual and other sensory information varies not just in terms of presence or absence. Sometimes they vary in intensity from weak to strong, and depending on the intensity of the sensation, we either detect it or not. Dogs and their powerful sense of smell illustrate the point well. Any adult who has spent time around dogs can tell that they smell things that are too faint for us humans. Dogs adopt a characteristic body posture, and their nostrils twitch excitedly when they smell something provocative. Not all levels of a sensory signal are detectable, and the ability to see, hear, or smell something can vary from individual to individual and species to species. As with seeing in the dark or understanding secret codes and games, when is it that children begin to understand that different agents have these different perceptual abilities and that they vary under different conditions?

In a preliminary exploration of these questions, Rebekah Richert and I asked three- to seven-year-olds to reason about various agents' perception under different sensory intensity levels.[7] We presented children with a seeing task, a hearing task, and a smelling task, all with the same basic logic. First, children were exposed to a target (the sight of a smiley face in yellow ink, the sound of a faint song, and the smell of peanut butter) while standing too far away to detect the target. Then the children were brought near so that they could see, hear, or smell the target. Finally, they were returned to the initial location (at which they could not see, hear, or smell the target) and asked whether a number of other agents would be able to see, hear, or smell the target from their position. Children answered about three puppets that they were asked to pretend were real: a girl named Maggie, a monkey, and a second special animal said to have extremely good ability to see (an eagle), hear (a fox), or smell (a dog). As in previous theory of mind studies, it was not until children were five years old that a majority consistently reported that Maggie and the monkey would not be able to see, hear, and smell the target from the initial position, whereas the special animals and God would. Indeed, until this age children tended to answer that they themselves could see the smiley face, hear the music, and smell the peanut butter even though they could not. Younger children tended to overattribute the ability to perceive to themselves, Maggie, and the animals. Before age five, they treated everyone like God.

These various perception tasks converge on two conclusions. First, years before they are eight, children are able to distinguish between God's (and some special animals') perceptual abilities and those of human beings (and other animals). Second, in some situations, children are capable of accurately predicting God's perspectives more accurately than that of fellow humans, at least until around age five. That is, in some respects, children who are still struggling to learn what their mother knows, sees, smells, and hears have already managed to understand what God knows, sees, smells, and hears in similar situations. It appears that the developmental course for accurately predicting other humans' mental states is longer—perhaps by two years—than accurately predicting God's mental states. Three-year-olds can get God, but not people, "right."

It may be possible that children perform so well on these tasks, theologically speaking, because of thorough indoctrination. They have been told time and time again that God knows everything, can see everything, hear everything, and smells everything; that God is a super-doer. They then successfully apply these platitudes about God to these novel tasks. It is not that understanding God is easier to the child's mind than reasoning about people in these same situations; it is just that they have had more instruction about God's attributes than about people. Maybe. But this explanation looks like a stretch to me: Is God's supersmelling really more prominent in their experiences than failures to see, hear, or smell something in other humans and themselves? Surely they would have learned that they cannot see or hear things from far away before learning that God has supersmelling. Research on divine immortality raises a similar problem for the indoctrination hypothesis.

Divine and Human Immortality

Recently a friend and colleague shared with me a story from his daughter's childhood. She had recently come into contact with human mortality through the death of a friend's mother. This event prompted many questions for the preschool-aged child and led to many discussions and explanations from parents about the permanency of death. Her father explained that death is the end. The body stops working and cannot start working again, much like a broken toy. To add to the finality of it all, the body is placed in the ground, and it turns into dirt. Nothing is left to fix. This idea of an irreparably broken toy or machine made sense, and the girl acquired an understanding of death. This relative ease for children to understand biological death is not surprising. But a specific application of the principle still concerned the little girl: What about Jesus? Is Jesus dead or alive? The father tried to explain the exceptionality of Jesus but got nowhere. Eventually the frustrated tyke conceded that such matters are simply "too hard for kids to understand."

Jesus dying, coming back to life, and then continuing eternally as an immortal being may very well be too hard for kids. The intrinsic immortality of Zeus, ghosts, Allah, and other beings, however, might not be all that difficult for kids to comprehend and believe. Like superknowledge and superperception, perhaps children begin with a default assumption that all intelligent beings—gods, humans, and some animals too—are immortal and have to modify this assumption when confronted with human and animal death, but need not change their thinking to accommodate immortal beings such as gods. Two experiments hint at this possibility.

How Children Think about Mortality

Marta Giménez-Dasí and collaborators asked Spanish three- through five-year-olds questions regarding the mortality of a friend versus God's mortality.[8] These questions concerned whether God and a particular friend (1) were alive at the time of dinosaurs, (2) were ever babies at some time in the past, (3) would grow old at some point in the future, and (4) would die at some point in the future. To create a mortality score for God or a friend, the experimenters gave 1 point for a "no" answer on the first question and 1 point for a "yes" answer on the remaining questions. For instance, a child who answered that his or her friend was not alive at the time of dinosaurs, was once a baby, would grow old someday, and would eventually die would have a mortality score of 4 for their friend. Answering that God was alive with the dinosaurs, was never a baby, would not grow old, and would never die would give that child a mortality score of 0 for God, indicating that God was immortal.

Not unlike in the knowledge tasks or the perception tasks, three-year-olds did not clearly distinguish between a friend's and God's mortality. Children did not clearly grant mortality to either, with the average mortality score for a friend equaling 2.0 and the average for God equaling 1.6. In contrast, the five-year-old children uniformly and accurately attributed mortality to a friend. Their average mortality score was 4.0. Again, as with the cracker box experiments, the five-year-olds did not anthropomorphize God or treat God the same as a human. Five-year-old children, for instance, regarded a friend as more likely to age and die than God. The average mortality score for God was still around 2.0, statistically indistinguishable from that of three-year-olds.

As in the studies looking at children's understanding of minds, this study demonstrates that five-year-olds need not treat God and people as the same. But does it suggest that three-year-olds are biased toward adopting godlike properties—in this case, immortality? The results were not clear. Three-year-old children had a tendency to reject the mortality answers for God, but their answers did not reach the level of consistency that would give us confidence. Rather, for God and for their friends, they seemed to be answering "at chance." That is, they might have had no ideas and were guessing. Perhaps. But perhaps what we saw in the three-year-olds when answering about their friends was a snapshot of going from assuming immortality to learning about their friends' mortality. In the case of God, maybe they did not quite give a convincing rejection of mortality because of the Christian context in which the study took place. Half of the children attended Christian schools, and all of the children were growing up in a cultural context in which God's immortality is complicated by God becoming fully human as Jesus of Nazareth. If some of these children understood "God" to include Jesus, a not unlikely scenario, they might have been inclined to answer that once upon a time God was indeed a little baby, and not only could but did die on Good Friday.

For these reasons, my student Emma Burdett and I replicated Giménez-Dasí's study with a sample of children who do not have the complicating problem of God becoming human. Burdett went to a town south of Jerusalem and talked to Jewish Israeli children in a place all-too-familiar with human death but lacking a God who was ever a baby or could possibly die. We asked the same four questions about God, the child's mother, and a friend. Our mortality score was based on three questions: whether God, the child's mother, or a friend can grow old, can die, and was ever a baby.[9] We divided our children into a younger group (three to four-and-a-half years old) and an older group (four-and-a-half to five years, eleven months). The younger children already distinguished between God and the two people and in the right way. Had children just guessed, they would have an average score of 1.5. The younger children, however, gave mother and friend average mortality scores just below 2. That is, they weakly understood that people are mortal. In contrast, their average mortality score for God was only around 0.8, indicating they already had a strong sense that God was immortal. When just looking at the question "Will God ever die?" twenty of the twenty-nine younger children said God would never die. Most children understood that God would not die. Children were not so sure about their mother or friend: twelve of twenty-nine said their mother would not die, and fifteen of twenty-nine (more than half) said their friend would not ever die. Older children (still only four or five years old) were essentially perfect in saying that God would not die but the humans would. What we have found in this study is evidence that children show an earlier appreciation of God's immortality than of human mortality, even in war-torn Israel. Perhaps stories about Jesus did create confusion in previous studies.

A slightly different wrinkle on immortality is that of the vampire. Vampires, as portrayed in popular entertainment these days, will never die of natural causes, but they can be violently killed. Is that how our Israeli children think about God? We asked, "Can God be killed?" Of the fifty-three children in the study, only four said yes, and two did not know. Even looking at the dozen children under age three and a half, nine said that God couldn't be killed, two said that God could be killed, and one did not know.

We can safely conclude that in line with the other studies described in previous chapters, five-year-old children need not anthropomorphize God with regard to mortality. Even for three-year-olds, it does not appear that children assume mortality and have to learn otherwise. If anything, it is the other way around: children have a modest bias toward attributing immortality and have to learn otherwise. In the case of God, if anything, three-year-olds showed a tendency to reject mortality. If further research bears out this tendency, we need not be surprised. The inevitability of biological death—let alone mental death—through natural causes may not be obvious in the day-to-day life of most children. And if minded beings need not be biological things, as I think the evidence shows regarding how babies reason about agents, then there is little impetus for children to think of a nonbiological God as mortal. They would have to learn that God is mortal through confused and anthropomorphic teaching when they could have just as easily, or even more easily, learned otherwise.

Immortality, Ghosts, and Afterlife Beliefs

When touring a wonderful historic residence in England, my wife and I had a peculiar experience. As we meandered through this sprawling palace, still home to the current duke, we examined the numerous portraits of the family and other esteemed guests and connections. Toward the end of one long hall of portraits, we came to a contemporary-looking portrait. My wife immediately exclaimed that she had seen this man earlier during our visit and that he was wearing the same unfortunate sports jacket as in the portrait. Naturally we assumed that this must be the current duke and my wife had chanced to spot him. Minutes later, we found an explanatory card that informed us that this was indeed the duke and master of this residence. Riddle solved? Not exactly. The dates listed showed that this duke had died four years previously.

In situations such as this, rare though they are, even those of us who do not believe that ghosts haunt the living can feel a momentary prickle on the backs of our necks. The possibility that a ghost has been fleetingly sighted temporarily trumps any reasoned explanation about a constructed false memory. An atheist colleague of mine shared a story about how he fleetingly believed a recently deceased relative tried to communicate with him. Another agnostic colleague feels uncomfortable speaking ill of the theories of a certain famous predecessor in his field when in his own office. Why? Because the office contains furniture that once belonged to the deceased. In short, something in us is ready to believe in ghosts whether we want to or not.[10]

A growing number of cognitive developmentalists believe that something about the way human minds develop appears to make us highly susceptible to believing that something in us persists after death and that something might continue to act in the present world.[11] Preliminary evidence of this susceptibility is the fact that beliefs in ghosts, saints, and ancestor spirits—all conceptions of formerly living people now dead but still able to think, feel, and act in ways that matter to the living—are among the most widespread type of supernatural belief. Even in places that claim to be atheistic, such as segments of Chinese and Japanese society, people still have rites for communicating with deceased ancestors. Standard Christian theology teaches that people have to be bodily resurrected to enjoy an afterlife and do not simply continue after death, and yet talk of ghosts and spirits of the dead communicating with the living continue unabated in Christian communities.[12]

In experiments and structured interviews, children also show an understanding of death that is very limited, as if it makes sense in only some situations or for only some aspects of humans.[13] For instance, developmental psychologist Jesse Bering and his collaborators tested the intuitions of Spanish children (five- to twelve-year-olds) regarding the afterlife (if any) of an anthropomorphized mouse.[14] Children watched a narrated puppet show in which Baby Mouse goes for walk in the woods, ending with, "Just then, he notices something strange. The bushes are moving! An alligator jumps out from behind the bushes and eats Baby Mouse. The alligator gobbles him all up. Baby Mouse is not alive anymore." After the children agreed that Baby Mouse was dead, they answered a series of questions relating back to the story. For instance, in the story, Baby Mouse grew hungry, so after his death, children were asked, "Do you think that Baby Mouse is still hungry now?" In the story, Baby Mouse was angry with his brother and wished he did not have one anymore, so after Baby Mouse died, the experimenter asked, "Do you think that Baby Mouse still wishes that he didn't have a brother now?" The Spanish children expressed more certainty that Baby Mouse would no longer need food or feel thirsty (now that he was dead) than that he would still want to go home or think about his brother. More clearly biological properties (like needing food) were easier for children to understand as ending with death than mental properties such as having desires or being able to think.

Several scholars point to the fact that death—instead of immortality—is complicated by the fact that it has physical, biological, and psychological components (at least). That is, physical bodies decompose and change with death; biological bodies stop growing, developing, breathing, needing food, and moving about; and minds stop thinking, perceiving, feeling, wanting, and being aware. If these different aspects of death require the coordination of different mental systems, difficulties may arise. What if one system, the one responsible for thinking about what a person wants and thinks (theory of mind), does not have a ready off-switch when the body stops?[15] Perhaps dead bodies then, even more so than mannequins, produce conflicted uncanny feelings and leave open the possibility that somehow the former occupant of the body is still watching, listening, thinking, and acting.

Exactly why believing in souls or spirits that survive death is so natural for children (and adults) is an area of active research and debate. A consensus has emerged that children are born believers in some kind of afterlife, but not on why this is.

Divine Goodness?

Little scientific research exists concerning how children regard the morality of God or other superhuman beings, and so I cannot confidently claim that children have any tendency to attribute moral goodness or badness to their gods generally. I suspect, however, that some sets of attributes more naturally congeal than others. If a god is superpowerful and superknowing, it may be more intuitive for it to be morally good as well. Relatively weak and dumb gods, in contrast, may be more prone to moral failure. To see why I have these suspicions requires a brief sketch of what appears to be emerging from scientific investigations of moral reasoning.

Intuitive Morality

Recent scientific research on moral reasoning, particularly research from an evolutionary perspective, is beginning to converge on the idea that from childhood, people have a basic set of moral instincts, a grammar, or intuitions.[16] These moral intuitions provide the skeleton for how people the world over think about what is right and wrong. For instance, we may all have an intuitive rule about not harming others without their consent, but this rule can be attenuated or amplified in different cultural settings. Teaching morality is a matter of heightening children's sensitivity to these moral intuitions, helping them to recognize these intuitions and apply them to novel situations, sometimes negotiating among competing rules. Evolutionary scientists have suggested that these moral instincts helped our ancestors cooperate and outcompete those without automatic moral intuitions.[17] These observations suggest that contrary to what some might argue, core morality does not come from a particular cultural belief system. All normally developing people have similar, basic moral intuitions, just as all people have basic intuitions about the properties of physical objects or how people's behaviors are shaped by their beliefs and desires.

One of these basic moral intuitions appears to be the belief that moral codes are absolute and unchangeable, whereas other norms are arbitrary and could be changed. So not putting your elbows on the dinner table is a norm that Mom or the president or God could overrule: "Go ahead. It is now okay to put your elbows on the table." But murdering your sister is never okay, and no one can make it okay. This view of morality—that its precepts are unchangeable, that some actions are intrinsically right or wrong—is sometimes called *moral realism*: there exist real moral codes that cannot be arbitrarily changed. Evolutionary anthropologists and psychologists suggest that people are naturally moral realists.

Though these observations about the naturalness of moral thinking have been championed recently by defenders of scientific atheism, they would come as no surprise to many religious believers.[18] In his treatise on education, *The Abolition of Man*, Christian scholar (and creator of the *The Chronicles of Narnia*) C. S. Lewis argued for universally converging moral intuitions, which he termed the Tao, and illustrated the Tao with quotations from philosophical and religious traditions from around the world.[19] Elsewhere he observed that people have strong moral realist intuitions:

> I believe we can learn something very important by listening to the kinds of things [people in disagreement] say. They say things like this: "How'd you like it if anyone did the same to you?" . . . Now what interests me about all these remarks is that the man who makes them is not merely saying that the other man's behaviour does not happen to please him. He is appealing to some kind of standard of behavior which he expects the other man to know about. And the other man very seldom replies: "To hell with your standard." Nearly always he tries to make out that what he has been doing does not really go against the standard, or that if it does there is some special excuse. He pretends there is some special reason in this particular case why the person who took the seat first should not keep it, or that things were quite different when he was

> given the bit of orange, or that something has turned up which lets him off keeping his promise.[20]

In this passage Lewis begins to make the case that people generally hold the perspective that some things are right and some things are wrong, and that's that. We do not argue those points. What we argue is whether the particulars of our behaviors are justifiable using these principles. When we are convinced of the moral legitimacy of our actions, we argue as if convincing others of the legitimacy is a matter of informing them of the circumstances accurately, not persuading them of a moral principle. If they know the facts (as we do), they will come to the same conclusion we do.

Morally Interested Gods

In discussing how we morally evaluate an illustrative sequence of actions, Pascal Boyer puts the situation this way:

> A disinterested third party who knew the facts [about stealing from a friend] would agree that stealing the money was shameful. . . . This at least is what we assume and why we invariably think that the best way to explain our behavior is to explain the actual facts. . . . Most family rows are extensive and generally futile attempts to get the other party to "see the facts as they really are"—that is, how you see them—and *by virtue of that* to share your moral judgements.[21]

Boyer argues that the reason that gods often get pulled into a moral system is this intuitive sense that knowing the facts is all that is needed to reach a moral judgment. Gods often have access to all the relevant information to make accurate moral judgments. With this access, gods know who has been naughty or nice. "So we intuitively assume that if an agent has full access to all the relevant information about the situation, that agent will immediately have access to the rightness or wrongness of the behavior."[22] As far as a given god has this relevant information, the god knows who is right and who is wrong.

At least superknowing gods, then, are easily pulled into moral considerations as arbitrators who might punish or reward. If something inexplicably bad happens to you, you might think, *I wonder what I did to deserve this?* Your neighbors might wonder, too, while assuming that someone capable of meting out punishment does know exactly what you did. A god with access to what you do—even in secret—is a reasonable source of punishment. Even relatively dumb gods often have this access to relevant information because of their invisibility (they could be watching right now) or their ability to hear or see things we want to keep hidden.

So from childhood, we assume that many actions are intrinsically good or bad. We hold intuitions that if someone knows the facts of the situation, that person will automatically know the goodness or badness of the action in question. Gods often do know the facts of the situation, so they know if someone has been good or bad. Fortune or misfortune can look (intuitively) like punishment or reward at the hand of a morally interested supernatural agent. These relationships make the idea of a morally interested god or spirit fairly natural and easy to understand and believe in once you have the idea of a god or spirit in place.

Morally Good Gods

But none of these factors mean that the gods must behave morally themselves or will be morally interested. Few theologies insist that the god or gods of their traditions exhibit consistent moral goodness, let alone moral perfection, as in Christianity, Islam, and Judaism. Far more deities morally resemble the immanently fallible Greek and Roman gods, who were prone to envy, lust, deceit, hatred, and pettiness. The much-beloved Hindu deity Krishna famously enjoyed a penchant for sexual play with countless dairymaids that would never be tolerated as moral behavior for a mortal. Among the gods, absolute moral goodness is exceptional, but so too is being a supreme, all-powerful, and all-knowing being instead of one of a host of roughly comparable limited and fallible superhumans. It may be that the attribute of being morally good fits comfortably only with some kinds of other divine superproperties.

The combination of being superknowing and superperceiving and supremely powerful may rest comfortably with being morally good as well—at least better than a fairly stupid and uninformed, weak god. If Boyer's and others' analyses are correct, we assume that a superknowing and superperceiving being will also know what is morally right in a given situation. In contrast, a being with limited knowledge and perception is likely to make mistakes in moral judgment as well. Perhaps, too, a being powerful enough to create a universe is powerful enough to satisfy any and all desires it might have without resorting to immoral behavior to satisfy its desires. A weak god may have to do immoral things to get what it wants, just as people resort to cheating and stealing because they cannot simply will what they want into being. A cosmic sovereign God that is all-knowing and all-perceiving has fewer reasons to behave immorally or be morally corrupt than a fallible god with regard to knowledge and power.

Of course, these observations do not constitute an argument for a superpowerful and superknowing God *actually* being morally good.[23] That one entails the other is not immediately evident. Rather, I offer these observations because on an intuitive level, we may find that superknowledge, superperception, and superpower in a divinity may mutually reinforce moral goodness when these attributes appear together. A testable prediction we might hazard is that supergods tend to be morally good and fallible gods tend to be morally dubious.

Sensitive Period for Belief?

That children have natural tendencies to acquire beliefs in some kind of god generally, and perhaps a superknowing, superperceiving, immortal, and good creator in particular, does not mean all children will inevitably become theists. These natural tendencies might get overridden by other factors, or perhaps children must be exposed to some kind of god concept early in life to capitalize on these early tendencies. The conceptual space for a god might be there in the first five years but then gradually shrink or reshape if the right god concept does not fill the space.

A large amount of research concerning how children acquire language suggests that people have what has been termed a *sensitive period* for language acquisition.[24] That is, during a certain age window in development, children need to be exposed to natural language in order to understand and speak a language at normal levels of competency. When they do live in such a language-rich environment, they take to language like fish to water. If they do not, language learning becomes much more difficult and may never reach full fluency. The sensitive period for language acquisition appears to be from birth until the start of puberty, though exposure during the first six years may be most important.[25] If a child is deprived of adequate language exposure during these early years, decades of subsequent exposure would likely still leave the child with severe language deficits.

Might children have something like a sensitive period for learning about God? Strong evidence has not been compiled to address this question, but several things that we know about cognitive development in childhood suggest the possibility of such a sensitive period. First, there seems to be something like a sensitive period in other areas of thinking that take natural tendencies and combine them with cultural peculiarities. For instance, some musical competencies seem to have a sensitive period. If children are exposed to music training during their early years, the time investment is more fruitful than later in life.[26] Related domains of thinking that have sensitive periods tend to be those that require a certain amount of environmental tuning. For instance, because different foods are available in different environments, deciding what is and is not disgusting cannot be hardwired into our brains. Some flexibility is needed, but total flexibility could lead to disaster, as in eating poisons or other dangerous substances. Consequently, what one finds disgusting takes some tuning up in early childhood. One- and two-year-olds can eat some disgusting things (as when my son, uh, cleaned up after the pet rabbit), but once something is determined to be disgusting in the third and fourth years, getting children or adults to change their minds is very difficult. So many Americans (but not as many East Asians) think that the idea of eating raw fish or live octopus is disgusting, and being exposed to the idea over years and years will not necessarily change their minds.

Learning about minds, including gods' minds, has some similarities to these domains in which a sensitive or critical period has been demonstrated to exist. Like language, reasoning about minds is a critical social competence necessary for human survival as social beings. Also like language, some variability exists from place to place in terms of exactly what properties of minds must be understood and attended to. The content of people's beliefs differs from place to place. Those objects and activities that they find valuable vary. Roles, offices, occupations, and social conventions vary. Further, when considering minds, people may not only master the types of minds that normal adults have, but also the kinds that young children have, or the aged, blind, deaf, or otherwise different people might have. Hunting or farming societies may have to learn about different animal minds. All of this potential variability in the environment, like all of the potential variability in languages, means that reasoning about other minds may require a certain amount of flexibility early in childhood. Like disgust, our reasoning about others' minds needs to be tuned to the kinds of minds that will be most important to us throughout our own life.

These parallels make me suspect that there might be something like a sensitive period for thinking about gods' minds. That is, learning that God is all-seeing, all-knowing, able to read minds, and attend to an infinite number of prayers all at once may seem difficult—too difficult for young children. But it could be that getting good at this sort of reasoning in childhood may prove easier than trying to learn it for the first time in adulthood. One advantage young children might have is that some of their thinking about all minds is naturally biased to assume superperception and superknowledge. Once we learn how knowledge and perception are restricted in humans, it might be harder to unlearn when applied to divinities. It could also be that children have more natural flexibility in reasoning about different sorts of minds than adults do, and so what looks, from an adult vantage point, as just too hard for kids might really be just too hard for adults but child's play for kids. Perhaps learning about God's superhuman properties is something like language learning. Five-year-olds can learn a second language in a matter of months, a nearly impossible feat for an adult. Maybe five-year-olds can learn to reason about many of God's superproperties more fluently than adults with the same amount of practice.

Building Blocks of Religion

Piaget wrote: "Here are the facts. The child in extreme youth is driven to endow its parents with all of those attributes which theological doctrines assign to their divinities—sanctity, supreme power, omniscience, eternity, and even ubiquity."[27]

Though I am not confident Piaget's list of divine attributes perfectly agrees with available evidence, his general notion strikes me as on the mark. Children "in extreme youth" think of agents—from humans to gods—as roughly godlike. Young children appear to have no particular difficulty in understanding a superknowing, superperceiving, superpowerful, natural-world-creating, and immortal God, and they might even have an inclination to assume these properties of all agents until they learn otherwise. In that sense, the basic, naive agent concept that children use to fashion their ideas about people, spirits, angels, devils, and gods might be more God shaped than human shaped, at least in the beginning. If true, this condition helps account for why children can readily acquire concepts of God and other supernatural beings: the basic building blocks are all in place before children learn about the particular gods of their cultural or religious group.

When conducting experiments of the sort described in this and the previous chapter, my student and collaborator Emma Burdett witnessed many telling (and often amusing) exchanges between children and adults, including their parents. One event from Oxford, England, captured the born believers thesis nicely. Before one session, a mother informed Burdett that her five-year-old son was introduced to the concept of God at school but that she was an atheist and they never spoke of religious matters at home. She said she was happy for her son to take part in the study, but had no idea what his responses would be given this scant exposure to the idea of God. During the session, this little boy responded in a theologically correct manner to the questions about God—answering that God would not die, would know what is in the cracker box, and the like. His mother sat flummoxed during the interview. Following completion of the tasks, she asked her son if he believed in God. His response, "Well, of course, Mum."

Burdett asked this mother if she would participate in the study, as we needed some adults for comparison. The mom agreed but explained that it was probably a waste of time because she would answer no to all questions about God because she was an atheist. True to her word, during the first task, this mother gave a "no" response for God. Her son, who was still in the room, rolled over on the floor to look his mother in the face and asked, "Mum, why are you saying 'no'? The answer should be YES!" He found his mother's obvious confusion a source of amusement throughout and kept laughing at her ridiculous answers. How could she get so wrong something that was so obvious to him?[28]

Part Two

The Implications

Six

Natural Religion

SCIENTIFIC RESEARCH on children's developing minds and supernatural beliefs suggests that children normally and rapidly acquire minds that facilitate belief in supernatural agents. Particularly in the first year after birth, children distinguish between agents and nonagents, understanding agents as able to move themselves in purposeful ways to pursue goals. They are keen to find agency around them, even given scant evidence. Not long after their first birthday, babies appear to understand that agents, but not natural forces or ordinary objects, can create order out of disorder. Before children start school, they see the natural world as purposefully designed—even in ways that religious parents would not teach or endorse. This tendency to see function and purpose, plus an understanding that purpose and order come from minded beings, makes children likely to see natural phenomena as intentionally created. Who is the creator? Children know people are not good candidates. It must have been a god.

Gods are not just humans with the ability to make mountains, trees, and butterflies, however. Early default assumptions about minded agents make it easy for children to understand gods as having full-access knowledge, superperception, superpower, immortality, and perhaps moral goodness. In fact, on some of these dimensions, children show the capability of reasoning in a theologically accurate way before being able to reason accurately about human beings on the same dimensions.

This collection of religious ideas is among the features of what I call *natural religion*. In this chapter, I describe natural religion and also how it deviates from theological beliefs. Though children have strong natural tendencies toward religion generally, these tendencies do not inevitably propel them toward any one religion. They still have a lot to learn.

Natural Religion

Children are born believers of what I call natural religion—parallel to the natural language that many linguists say children's minds are naturally inclined toward understanding. English, Hindi, Mandarin, Spanish, Swahili, Yukatec, and the other languages of the world are derivations and elaborations of this natural language. Similarly, Christianity, Hinduism, Islam, Jainism, Judaism, Mormonism, Sikhism, and other tribal and world religions are derivations and elaborations of natural religion. Natural language more firmly constrains the world's languages than natural religion does the world's religions, but natural religion still provides anchor points for the world's religions from which they will have difficulty straying. The study of natural religion remains in its infancy (especially as compared to the study of natural language), but we may tentatively project some of the features of natural religion.

Research into children's acquisition of religious ideas and cross-cultural comparisons suggest that natural religion includes several assumptions:

- Superhuman beings with thoughts, wants, perspectives, and emotions exist.
- Elements of the natural world such as rocks, trees, mountains, and animals are purposefully and intentionally designed by some kind of superhuman being(s), who must therefore have superhuman power.
- Superhuman beings generally know things that humans do not (they can be superknowing or superperceiving, or both), perhaps particularly things that are important for human relations.
- Superhuman beings may be invisible and immortal, but they are not outside space and time.
- Superhuman beings have character, good or bad.
- Like humans, superhuman beings have free will and can and do interact with people, sometimes rewarding and sometimes punishing them.

- Moral norms are unchangeable, even by superhumans.
- People may continue to exist without their earthly bodies after death.

This natural religion then gets specified, amplified, or even contradicted in particular cultural settings—what we often call *theology*—not unlike how we learn the particulars of our native language. For instance, in a Christian cultural context, the features of natural religion might be theologically elaborated in the following way (elaborations in italics):

- Superhuman beings with thoughts, wants, perspectives, and emotions exist; *they consist of a supreme God and lesser beings such as angels and devils.*
- Elements of the natural world such as rocks, trees, mountains, and animals are purposefully and intentionally designed by some kind of superhuman being(s), who must therefore have superhuman power. *The natural world was created by a single God.*
- Superhuman beings generally know things that humans do not (they can be superknowing or superperceiving, or both), perhaps particularly things that are important for human relations. *God knows absolutely everything that can be known and nothing can escape God's attention, whereas other superhuman beings, such as angels or Satan, are superhuman in knowledge but still limited.*
- Superhuman beings may be invisible and immortal, but they are not outside space and time. *God and the*

other superhuman beings are typically invisible but may become visible. God is immortal and is eternal. Other superhuman beings are not mortal in the biological sense but are immortal only if God allows them to be, and they do exist in time.

- Superhuman beings have character, good or bad. *God is perfectly and unchangingly good; the other superhuman beings may be good or bad.*
- Like humans, superhuman beings have free will and can and do interact with people, sometimes rewarding and sometimes punishing or extracting revenge. *God sometimes directly punishes wrongdoing.*
- Moral norms are unchangeable, even by superhumans. *Moral norms are an expression of God's infinite character and as such are unchangeable.*
- People may continue to exist without their earthly bodies after death. *God may choose to resurrect people after death to continue existing in heaven.*

Note that these illustrative elaborations are harder to understand and more cumbersome to express than the jumping-off point supplied by natural religion—and I tried to keep them simple. The more complex that theological ideas are—that is, the more they deviate from the ordinary cognition that undergirds natural religion—the more effort that will be required to teach them and maintain them, a point I return to and illustrate.

Given the relative difficulty of grasping particular theological claims as opposed to general natural religion, it is no surprise that children, who do not have the same intellectual resources as adolescents or adults, often get theologies wrong. Though the naturally developing wiring of human brains typically produces minds ready and willing to believe in gods as intelligent, intentional beings that designed the natural world, perhaps also having infallible beliefs, possessing superperception and superpower, and being immortal, children do not have the sophisticated, fleshed-out theologies some adults acquire. Let's face it: some ideas about gods are just plain hard to understand, let alone believe. Piaget got a lot right in regarding many ideas as beyond the grasp of young children. Some theological ideas escape even adults' comprehension.

Take the doctrine of the Trinity, for instance. Traditional Christianity maintains that there is only one God and God consists of three "persons": God the Father, God the Son, and God the Holy Spirit. The careful formulation of this doctrine and what it means and does not mean has been the subject of numerous church councils and theological treatises spanning more than a millennium. I suspect that for most of the world's 2 billion Christians, this doctrine remains a hard-to-understand mystery but not the sort of thing to spend too much time fretting about. Many critics of Christianity conclude (a bit prematurely, I think) that the doctrine is absolutely incomprehensible and therefore nonsense.[1] My point is not that the cognitively unnatural or conceptually cumbersome concepts of theology are wrong or right, only that they take more time and effort to acquire, and they certainly fall outside children's natural predilections. Culturally special conditions need to be in place for these kinds of ideas to successfully spread.

Cognitive biases and tendencies, born out of maturationally natural systems and not cultural particularities, make children born believers. Nevertheless, it would be misleading not to point out that a gap still exists between what kinds of supernatural beliefs come naturally to children and what kind of theological beliefs are endorsed and promoted by adult theologians. Children may be born believers, but it would be a stretch to call them born theologians. Both the kinds of ideas that theologians develop and the intellectual practice of doing theology may be relatively unnatural.

Less Natural Religion

In the following sections, I offer theological ideas that I regard as fairly unnatural and not what children are prone to adopt. I then take a brief detour into the difference between religion and theology to make the point that though children are born believers in religion, they are not born theologians.

Strict Monotheism

Remember Zeus, Apollo, Hermes, and Venus from grade school? Or perhaps Anubis, Ra, Osiris, and Isis? The ancient Aztecs, Babylonians, Egyptians, Greeks, Incas, and Romans all had a large range of gods with divisions of labor: sun gods, fertility gods, gods of the dead, and so forth. Similarly, Hinduism features perhaps hundreds of gods and goddesses, and the Maya of today have a large pantheon of gods. Given the historical and cross-cultural evidence, it would be hard to argue that belief in only one god is more "natural" than multiple gods.[2]

Christianity, Islam, and Judaism are known as the great monotheisms, all insisting that there is only one God, but even in these traditions, people believe in numerous supernatural beings, including angels, devils, saints, and ghosts. In a certain respect, believing in numerous superhuman agents appears to be the most natural type of belief system. We have no evidence that suggests children are born monotheists in a strict sense. Packaging all ambiguous experiences from the hypersensitive agency detection device—including episodes of great fortune and misfortune and the apparent detection of the recently deceased—together with accounting for design and purpose in the natural world as attributable to one being might require too much abstraction and intellectual nimbleness to be terribly common.

Nontemporality

Many Muslim, Jewish, and Christian theologians teach that God is outside time, or nontemporal. This may be one way to make sense of scriptural passages such as 2 Peter 3:8, which reads: "But, beloved, do not forget this one thing, that with the Lord one day is as a thousand years, and a thousand years as one day." God being outside time also features in some explanations for how it is that God can know our future without taking away our free will. But what does it mean for God to be outside time? I have trouble imagining what that might be like or how to talk about a being outside time. Taken seriously it means that there is something peculiar about saying that "*when* I pray, God hears me" or that "*in the past* God spoke to Abraham." I would be very surprised indeed if young children have no problem with God's nontemporality.

Nonspatiality

The idea of a being that does not have any location in space—about whom we cannot really say it is here, there, anywhere, or everywhere—strikes me as extremely unlikely to be readily embraced by children. As with nontemporality, this nonspatiality figures in many theologies. If God is not a material being, it is misleading to say God is *in* heaven or God is every*where.* These expressions amount to metaphors because a being with no location is too hard to handle—for children or adults.

Unlimited Attention

It is one thing to be able to see in a darkened box or through barriers or to hear something from across the universe. These properties of a divinity appear to be among those children are ready to make sense of and accept. It is another thing to pay attention to what is in every darkened box everywhere and every sound emitted from every corner of the universe. Here I am pointing to the difference between being able to see and actually watching, or being able to hear and actively listening. A more common expression of this unlimited attention problem is the idea that God watches you and everyone else all the time or that God listens to everyone's prayers from all over the world at the same time.[3]

We have no reason to believe that children or adults find the property of unlimited attention the least bit natural or intuitive. Our own limits on attention and inability to imagine what it would be like to simultaneously hear everyone's thoughts or watch everyone's activities makes the idea of any being having such a property hard to readily comprehend. I suspect that when confronted with the idea of a being that can pay attention to everything, we are prone to recast the property as simply knowing everything. Can God pay attention to everything at once? What would that mean? It would mean God knows everything. God knowing everything, or having access to any scene you would like to mention, is not so hard to understand.

Grace—Maybe Easy for Kids But Not for Adults?

A theological concept that I have struggled with regarding its relative naturalness for children to grasp is grace. Grace is sometimes defined as unmerited favor or not getting the punishment we deserve. In Christian theology, grace captures the idea that salvation is not earned or deserved but is a free gift from God to those who receive it. Think of a Christmas present from Grandma when you were a child. You didn't earn it, and there is no expectation of reciprocating—no quid pro quo. To get it, all you have to do is say "thank you" and unwrap it. Many Christians say God's grace is like this present: receiving it and expressing gratitude is the only reasonable response.[4]

I use the image of a Christmas present from Grandma because it illustrates the potential difference between children's and adults' reactions to the notion of grace. In public question-and-answer sessions after lectures and addresses, I have suggested that the Christian doctrine of grace is not terribly natural. Adults, at least, through the ages seem compelled to add stipulations to the gift of salvation: that you have to behave yourself, go to church, wear trousers if male and dresses if female (instead of traditional ethnic attire), read the Bible frequently, pray daily, and so forth. That so many preachers include sermons on grace in their regular offerings of topics, month after month, year after year, attests that the message just is not sticking. People seem to have a deeply ingrained sense of fair exchange practices. If you give me something, I am obligated to give you something of comparable value. If I do not reciprocate, I am in your debt. Being in someone else's debt is uncomfortable, so we try to settle the account. Worse still is when someone else receives grace instead of justice. Theologian Donald McCullough puts the situation thus:

> Grace set to music is one thing, but what about grace itself? What about grace as an idea? Grace as an act? Grace as a force? We don't like it. . . . To be sure, we appreciate brief encounters with it, such as when we forget to pay the health insurance premium and we're told not to worry because there is a "grace period." . . . We're thankful for these minor reprieves, for Grace Lite. But if the real thing happens, if we're seized by a full-bodied, take-no-prisoners grace, we have far more ambivalent feelings. When a muscled arm of mercy lifts us by the scruff of the neck and sets us in a new place, a better place we neither earned nor deserved, we're likely to protest that, given time, we could have gotten ourselves there, thank you very much, and without the rough treatment. Even worse, if grace happens to someone else, someone we *know* doesn't deserve it, someone we can't stand, then we don't want to hear about grace, let alone see it in operation. In such circumstances, grace seems more like a miscarriage of justice.[5]

Evolutionary psychologists Leda Cosmides and John Tooby have talked about how sensitive people are to social exchange rules and have argued that this sensitivity is an evolved cognitive capacity.[6] Even the feeling of gratitude may be tough for us. If someone pays us an enormously generous kindness, we may feel embarrassed, guilty, or indebted instead of grateful.[7] So no matter how many times preachers tell us, "God doesn't need anything from you," or "The price Jesus paid is too big for any mortal to reciprocate," or the like, we just cannot seem to shake that nagging feeling that God wants something from us in exchange for salvation. These sorts of considerations have led me to think sometimes that grace is counterintuitive to the way people think.[8]

But maybe children do not have the same problems with grace and gratitude that many adults have. Unlike adults who might have a deep sense of obligation and giving like-for-like, children might have no such feelings. Children, especially very young ones, do not have the resources or ability to reciprocate in a tit-for-tat fashion with others, and they are not often embarrassed by others giving them gifts or doing things for them. Pride does not get in the way. If Grandma gives a child a gift of a trip to Disneyland, he is not going to feel uncomfortable or wonder about how he will repay the gift. In his excitement, he might need reminding to say "thank you," but it isn't because he is not thankful or gushing with gratitude. (When Aunt Mabel gave him the five-pack of argyle socks, that was when he was ungrateful and needed reminding to be courteous anyway.) If children did have the same hang-ups with allowing others to be gracious toward them—giving what was not earned or strictly deserved—they would be riddled with angst and as soon as they were teenagers and began collecting their first paychecks, they would immediately give it to their parents with penitent promises to pay back their immeasurable debts as soon as possible.

These observations, speculative as they are, make me wonder whether children might be better able to accept grace from God too—not feeling the need to earn salvation or add stipulations and conditions. Could this easy acceptance of grace be what Jesus meant by saying, "Let the children come to me, and do not hinder them, for the kingdom of heaven belongs to such as these"[9]?

Animism

The examples above are all theological ideas that deviate from natural religion. We may associate antimism with traditional religions and assume that it is part of natural religion. Actually animism too is an intellectual or theological elaboration of natural religion. Some traditional and new age belief systems hold that rocks, mountains, and streams have spirits or life forces or that nonanimals have consciousness. These animistic ideas are sometimes thought to be among the earliest of religious beliefs, in part because of the common but misguided notion that children are confused as to what counts as a living thing and what does not, and that children readily treat nonconscious beings as conscious, intentional agents. The reasoning seems to be that if children have to learn what is animate and what is not, then so too did earlier peoples. Apart from the danger in assuming past peoples were like contemporary children, the evidence that children are naturally animists is shaky at best.

In Chapter 3 I shared that when my daughter was two years old, she used to pick up large earthworms in the yard and then carry them around, talking to them and cooing at them as if they were babies. She used to do the same with a feather duster. We even found her putting cans of vegetables in her toy stroller and acting as if the cans were babies. These behaviors are the sorts that we might take as evidence that young children do not know the difference between people and earthworms or between living and nonliving things. But is it really good evidence? This same child who treated the feather duster as a baby also used the feather duster to dust her room and knew to return worms to the ground or they would "die," but she never tried to use the earthworms or canned goods to clean her room or buried the canned goods or feather duster in fear they might die. Even two-year-olds are not stupid. While they may pretend that objects like dolls or feather dusters are alive or conscious, they know that they are not.

Plenty of experimental evidence demonstrates that preschool-aged children importantly discriminate between living and nonliving things. For instance, within the first two or three years of age, children know that objects such as plants and balls must be touched in order to move, whereas animate beings such as people can interact at a distance.[10] By age five, children seem to have a broad range of biological expectations for living things that do not apply to rocks, balls, or other nonliving things. For example, living things are thought to have an internal life force that makes them grow and move and natural inside parts, whereas machines have artificial parts. Living things cannot be changed from one category to another by external modifications (such as making a skunk look like a raccoon), whereas tools and other human-made things can. Animals have baby animals and move to eat and survive, but rocks and sticks do not, and living things have parts that do things for them, whereas human-made objects have parts that do things for the people that use them.[11] In fact, there is reason to believe that young children underascribe being alive (such as to plants, fungi, and nonmoving primitive animals) more than then overascribe, not recognizing that lichens, mushrooms, molds, and trees are alive as early as they understand that dogs, birds, and snails are.[12] The idea that children and, hence, early peoples assume that all objects are alive or conscious simply does not stand up to scientific evidence. There is no reason to believe, therefore, that children are intuitively drawn to animism of the sort we see in religious traditions.

Some aspects of animism, as seen in many belief systems of adults, however, might not be all that unnatural. One key component of natural folk biology—the idea that living things have an animating life force or vitality—may provide the nonreflective timber from which various reflective cultural beliefs are built, such as animating spirits, life forces, the Chinese *qi*, and other beliefs about unseen internal energies (spirits) that animate humans (and sometimes other things). Judaism and Christianity share the notion that the breath or spirit of God is an animating and energizing force transforming inert matter into living things, or making people and animals more or less lively and energetic. Perhaps all of these notions of spirit as an animating force might find their intuitive appeal in natural cognitive thought concerning biological thought. What is slightly unnatural and counterintuitive about animism is the suggestion that even rocks and trees are as comparably spirited as people or animals.

Not Born to Believe a Particular Theological Tradition

Throughout this book, I have tried to make the case that recently acquired scientific evidence suggests that children have a naturally developing receptivity to many core religious beliefs, particularly beliefs about the existence of supernatural beings. Given little environmental encouragement, they become believers in superhuman agency. But this natural receptivity to religious ideas is limited. Many theological ideas, of the sort that religious specialists develop and many believers affirm as part of historic creeds, do not number with those children are biased to acquire. Rather, these theological beliefs (such as nontemporality, nonspatiality, and the like) are conceptually difficult for children (and adults) and require special cultural scaffolding to spread effectively. In this regard, theological ideas share much in common with other ideas generated reflectively in special cultural conditions such as those found in modern science.

One implication of this limitation for the born believers thesis is that children are not "born believers" in any specific religious or theological tradition. After public lectures arguing for the born believers thesis, I received numerous e-mails and found blogs from Muslims claiming that the born believers thesis is similar to a standard teaching in Islam. Here is an illustrative note:

> Hallo, I read the article of Dr Barrett about believing in GOD. As a Muslim I believe in GOD and I know a lot about what Dr Barrett explained that every human is born believing in GOD naturally. Our Prophet Mohammed told us before more than 1400 years ago when said each fetus born believing in GOD naturally, and his father or mother makes him differ.

I appreciate this affirmation, but the evidence to date supports only the naturalness of belief in a mighty creator God, not that children are born to believe in orthodox Muslim, Jewish, or Christian theology—or any other for that matter. They may be biased in a general sense toward some religions over others—perhaps this is one reason that some religions spread more widely than others—but I know of no religious tradition that perfectly maps onto those ideas that children are inclined to believe.

Some who have a romantic view of children being uncorrupted and, hence, possessing access to life's truths most directly, might try to build a theology based on children's natural religious proclivities. Those who regard children as basically stupid might regard the natural receptivity of children to certain religious beliefs as grounds for rejecting them. I regard both impulses as misguided, a discussion I present in the next chapter.

Religion Is Not Theology

The cognitive science of religion, in which I participate, typically distinguishes between religious and theological thought. There seems to be a difference between what people tend to believe in an automatic, day-to-day sort of way and what they believe when they stop to reflect and systematically figure out what they do and do not believe. Some ideas, such as the particular sense in which Krishna might be Vishnu but not exactly the same, or the Christian God might be three persons at one time, or how exactly karma works, or what precisely happens to a Mormon after death are the sorts of issues that theologians rigorously ponder and argue about in hopes of getting things right. Theologians have spent, and continue to spend, enormous amounts of attention and energy on trying to work out the reasonableness of different propositions regarding God (or gods) and related matters. They draw on philosophy, science, textual studies, linguistics, and historical considerations to reach their conclusions. Such intellectual activities do not often characterize the behavior of devoutly religious adults, let alone children. Most individual believers do not engage in such theological exercises but are content to live religiously. To be religious is not to be a theologian, or vice versa.

A similar distinction between folk ideas and formal, reflective beliefs appears in other domains. For instance, by four years old, children have a sense for the basic grammar of their native language. If they are English speakers, they know that "the dog likes to eat cucumbers" is grammatical (even if strange), but "the likes to cucumber eat dog" is nonsense.

This folk language capacity, however, is distinguishable from the sorts of reflective knowledge about language use that adult specialists acquire in studying language. A linguist might be able to tell us more precisely the relationship among various parts of speech in English and why it is that "the dog likes to eat cucumbers" is better formed than "the dog, cucumbers to eat, he likes" (as *Star Wars'* Yoda might say), and various other specialized knowledge—none of which is necessary to successfully use the English language to talk on the telephone, order food at a restaurant, or share gossip by the watercooler. Regarding language, then, we can distinguish between folk knowledge of language and linguistics.

Although developmental psychologists sometimes draw parallels between the way children learn about the natural world and the way scientists do, a large difference exists between folk understandings of the natural world and scientific ones. Science, linguistics, and theology, on the one hand, and folk knowledge, language, and religion, on the other, differ in terms of degree of conscious reflection, effort, and commonness. The first group contains examples of relatively rarified thought that not all people engage in or care about. Because these kinds of thought take time and effort—they do not come naturally to us—they have not developed in all cultural contexts, let alone in all individuals. These observations imply that you can have knowledge about the natural world, language, or religion without having much, if any, knowledge that might be called science, linguistics, or theology.[13]

From Natural Religion to Theological Diversity

The picture of religious development arising from the research presented here (among many other studies) is that children are naturally drawn to some basic religious ideas and related practices (natural religion), and then the meat of a religious and theological tradition as taught by parents grows on this skeleton. It is this later cultural elaboration that provides the bewildering variety of theological beliefs that we see when surveying the world's belief systems.

Ideas about gods from around the world come in dizzying diversity. Some have animal forms, some have human forms, and some have no form at all. Some are all-knowing; some know about the same amount or kinds of things as people. Some are morally upstanding, and others are cruel. Some gods possess superpowers, and others have very little power. And sometimes a god is regarded as a kind of unknowable, indescribable otherness.

Among the Mali Baining of Papua New Guinea, the forest spirits, or *sega*, are strikingly human-like beings. Anthropologist Harvey Whitehouse observed that the *sega* are so human-like in appearance that they are hard to distinguish from people. "*Sega* are thought to look like humans, although few people have actually seen any, except in dreams. People cannot readily explain how they recognized what they had seen as a supernatural being, rather than as a mortal stranger."[14] The *sega*, however, are usually invisible to humans, their presence made known through fortune or misfortune:

> *Sega* are not offended by moral transgressions injurious to humans, and only punish unwanted interference in their own affairs. The problem is that, unlike dangers in the forest that can be seen (e.g., pythons, nettles, sharp objects), *sega* cannot usually be identified and avoided, and the fact that they have been disturbed or provoked only becomes apparent later, when misfortune strikes.[15]

In contrast to the humanness of the Baining *sega*, the great monotheisms offer strikingly complex, abstract views of God, often stressing how different God is from people and how incomplete or inadequate our understanding of God can be. For instance, Islamic theologian Mohammad Zia Ullah wrote:

> God is infinite, pervasive, and man finite and limited to a locality. Man cannot comprehend God as he can other things. . . . God is without limits, without dimensions. . . . How can a limitless, infinite being be contained in the mind of a limited being like man?[16]

Comparably, Christian theologian Gordon Spykman, in discussing the biblical view of God, explains that

> on this view, God and the world are two uniquely distinct realities. The difference between them is not merely quantitative but qualitative. God is not simply more of what we are. There is an essential discontinuity, not just a share of difference, nor a gradual more-or-less distinction, as though God has only a "running head start" on

> us. God is absolutely sovereign, "the Other," not simply "Another."[17]

For these concepts of God, our best, most dedicated, and rigorous intellectual exploration can only scratch the surface of what God really is.

Sometimes in a single belief system, gods can take on different sorts of forms, ranging from fairly human-like to abstract. In anthropologist Emma Cohen's descriptions of the Afro-Brazilian spiritualists' beliefs she studied in northern Brazil, we see great diversity in how the *orixás* (spirits or gods) are conceived. As the spirits of deceased ancestors living in a spirit world largely identical to our own, many *orixás* look and act just like humans. "The gods that feature in these stories display very humanlike demeanors in their social dealings with one another, acting on the basis of their desires and whims. Jealousy, vengeance, bad blood, and trickery color the conniving between the characters of this Elysian soap opera."[18] One of Cohen's informants saw the *orixás* as including some that are human-like in form and behavior and some as more abstract:

> But I saw the form of the *orixás* to be very black human beings . . . some strong and other things, some bodies were model-like and others not. . . . I discovered that the *orixás* are also forces of nature, and they are also present with us at all times. I discovered that there are *orixás* that participated in the creation of the world, and for having done heroic feats, they had the privilege of becoming *orixás*—kings, queens, founders of cities. So, really the *orixá* is everything that we can see and that we can feel.

> . . . I also discovered that they have qualities and flaws—some are very similar to human beings in that they both get things right and commit errors.[19]

Human beings or forces of nature? "The Other"? Can adults believe in just about any sort of god they can dream up? Perhaps, the sky's the limit—or maybe not even the sky.

What Adults Believe When They Are Not Watching Themselves

One reason to investigate childhood religion is that we are interested in children in their own right. Another reason for turning our gaze to the early years is that we can sometimes get insights into the adult situation by looking at from where we came. This early developing natural religion is not simply something we replace with adult theologies. Rather, it continues to exert an influence on how we think and act religiously throughout life. Adults are not impervious to the influences of maturationally natural cognition as manifest in natural religion.

Perhaps for any given person at any one moment in time, we cannot easily limit human religious imagination. In general, however, adults are not unfettered. Though someone might sincerely believe that God is the unknowable, indescribable, wholly transcendent otherness that is reality outside time and existing in an infinite number of dimensions of reality, this concept of God hurts my head. I cannot make sense of it and consequently cannot really use it (at least in a practical sense), even if I wanted to. Maybe I am unusually dense, but I have a feeling that I am not alone in having genuine limitations on what I can understand and believe. Fortunately, too, I have some experimental evidence that backs up my hunch.

Collaborators and I conducted a series of experiments in which we contrasted what adults said they believed about God with how they thought about God in a less reflective, real-time situation—understanding stories.[20] The inspiration for these studies was the sense that adults, even theologically savvy ones, do not always seem to use these theologically sophisticated theological concepts all the time. Spend time around believers in God—God with no location, no human-like form, outside time—and you will hear them say things like, "When I went through that ordeal, I felt like the Lord was walking right beside me," or "Sometimes when I pray, I imagine myself embraced in God's arms." This sort of language suggests a conception of God far more human-like and far less abstract than what theologians sometimes produce and believers affirm. Perhaps, however, these are just relational metaphors, figures of speech meant to convey a feeling or image but not really indicative of how people really think about God. Determining how people think about God in real-time ordinary situations only on the basis of the language people use is tricky. For this reason, collaborators and I conducted several experiments to try to clarify matters.

Stories present a wonderful opportunity for uncovering what readers or listeners bring to the stories themselves, a way of indirectly tapping their thoughts and intuitions. The reason stories are so effective in this way is that they are always incomplete. They have gaps that we often do not even notice. No storyteller can relay all the details in a story. If I say, "Once there was a girl named Cinderella who lived with her wicked stepmother and two stepsisters," you immediately (and probably unknowingly) fill in all of the details you know about girls, wickedness, stepmothers, stepsisters, and so on. If people did not automatically fill in all that they know, the story would never get off the ground. The storyteller would have to say things such as, "Once there was a girl—oh, a girl is the female of the human species usually before reaching reproductive and physical maturity and characterized by a diminutive stature relative to the mature form of both the male and female adult human . . ." If you've ever told a story that is not entirely age appropriate to an inquisitive child, you know what I mean. It takes a lot of background information to understand any story.

Cognitive psychologists—those who study memory and thought—have shown that adults easily and automatically fill in the gaps in stories to the point that they actually misremember what was in the original story as being more complete than it was in some ways. That is, we use our current knowledge to elaborate or distort what is presented.

The most famous psychological research demonstrating these sorts of intrusion errors and related distortions was that of Frederic Bartlett.[21] Bartlett presented British university students with an unfamiliar Native American story, "The War of the Ghosts," and then asked them to retell it to another person, who retold it to another person, who retold it to another person, and so on—like a game of telephone. Among Bartlett's many observations about how the story was remembered was that there was a strong tendency for hearers to distort or insert information that fit their own preconceptions. So people in a *canoe* became people in a *rowboat*. Less comfortable or familiar ideas were misremembered as more comfortable or familiar ideas. In a follow-up set of experiments, Melanie Nyhof and I replicated Bartlett's study presenting other Native American stories to non–Native American American students.[22] Again, less familiar things (such as buffalo chips) were misremembered as more familiar things (wood chips). Other researchers have documented how context-related and other knowledge can be misremembered as having been in an account.[23] For instance, in one set of experiments, participants heard: "John was trying to fix the birdhouse. He was pounding the nail when his father came out to watch him and to help him do the work." Subsequently a large portion of the listeners confidently agreed that the following sentence was one of the sentences they heard: "John was using the hammer to fix the birdhouse when his father came out to watch him and to help him do the work." Notice that in the original sentences, no hammer was mentioned. Listeners who know that nails are typically pounded with hammers naturally inserted the presence of a hammer in their memory of the sentences.

My collaborators and I took advantage of this tendency for the hearer or reader of a story to automatically fill in the gaps using their own ideas to create an indirect measure for people's ideas about God.[24] We constructed a number of stories that included God as a character but carefully left gaps for our audience to fill. For instance, one story read:

> A boy was swimming alone in a swift and rocky river. The boy got his left leg caught between two large, gray rocks and couldn't get out. Branches of trees kept bumping into him as they hurried past. He thought he was going

> to drown and so he began to struggle and pray. Though God was answering another prayer in another part of the world when the boy started praying, before long God responded by pushing one of the rocks so the boy could get his leg out. The boy struggled to the river bank and fell over exhausted.[25]

The questions we sought to answer with stories such as this one were, Just what kind of idea of God do listeners or readers use to make sense of the story? Is it the same as what they say they believe about God?

To answer these questions, we asked memory questions after our participating adults listened to the stories. (We encouraged the listeners to use their own concept of God as much as possible and directly asked them a number of questions about what properties they thought God has.) Because we were particularly looking for intrusion errors—instances when people remembered something that was not in the story—we asked: "Which of the following pieces of information were included in the story?" Listeners answered simply yes if they remembered it being in the story or no if they did not remember it. We assured our participants that the wording did not have to be exact. Some of the items checked for general memory of the story—for instance, "The boy was swimming alone." (Do not look back at the passage. Yes or no?) The other items checked for intrusion errors related to ideas about God. For instance, one item read, "God had just finished answering another prayer when God helped the boy." (Was that in the story? Yes or no?)

What we found using stories and questions like these is that listeners' intrusion errors revealed that they used a very human-like, or anthropomorphic, understanding of God to make sense of the stories. Did God just finish answering another prayer when God helped the boy? "Yes" was the most common answer. But look again at the story and remember that "God" means someone that can do any number of things anywhere at the same time. Does the story really say that God *finished* doing one thing and *then* did another? Couldn't the story be understood as God *continuing* to answer the prayer in another part of the world while beginning to help the boy in the river? Sure it could. When we substituted space aliens with superpowers for God in the same stories with a different group of adults, they did not make these kinds of intrusion errors at any higher rate than other kinds of mistakes. But there is something hard about understanding the story with an all-present, all-powerful, nonanthropomorphic God. At least when trying to use our ideas of God in these kinds of tasks, a very human-like concept of God seems easier, more natural.

We conducted these experiments with adults living in the United States from many different religious backgrounds and commitments. Some participants did not believe in God and had to be asked to use their ideas about the God in which they do not believe. (This is not as strange as it sounds. You might not believe in dragons, but you probably have a lot of ideas about what a dragon is.) Across all groups—believers or nonbelievers, Christians or Jews, Catholics or Protestants—everyone showed the same pattern of intrusion errors. They understood God (but not the carefully described space aliens) as human-like in the stories, but denied that they believed God to be human-like in the same ways when asked directly. In the stories, they incorrectly remembered God as being in one place, but when asked directly, they claimed God was everywhere or nowhere. In the stories, they incorrectly remembered God as having to do one thing at a time, but claimed God could do any number of things at once when asked directly. The God in the stories could be interrupted, could have vision blocked, and could fail to hear something because of competing noise. Participants explicitly denied all of these limitations on God.

Concerned that listening to the stories and the accompanying comprehension questions somehow unfairly pushed a more anthropomorphic God concept than people might otherwise use, we tried a version of the task in which adults read the stories themselves and then answered the questions. Same results. We also tried a task in which adults read some stories and then wrote out their own paraphrased versions. Sure enough, the anthropomorphic intrusion errors cropped up again.

Once after I presented this in a public lecture, a fellow psychologist who happened to be the wife of an Orthodox rabbi asked me, "What did you expect?" How else would people think about God under these conditions? Of course we need to use a simpler, more familiar idea of God sometimes.

Exactly. The reason I present these experimental results is that they demonstrate that in a sense, adults may actually have two or more different sets of ideas about God: one set is the fancier theological set about an all-present, all-knowing, and radically different kind of being that comes up in reflective situations; the other is the one that looks much more like a human and is easier to use in real-time situations. Ideas that deviate too far from our natural conceptual tendencies are difficult to use.

To learn whether these experimental results were peculiar to American adults. I traveled to India, a wonderful place of comparison as there are so many different deities in Hinduism, the religion of the large majority of Indians. These gods take a huge variety of forms and are often graphically depicted in drawings, paintings, and sculptures. Nevertheless, Hindu theology teaches a more abstract version of the ultimate being, Brahman. God is regarded as in and through everything. On one hand, they have more concrete gods than in Christianity and, on the other, more abstract divinity than in Christianity. Perhaps this observation by itself demonstrates a dual-level representation of gods in Hinduism: the abstract and the anthropomorphic side by side.

I replicated the same story–comprehension study with just a few modifications.[26] First, I used only the version of the task in which participants read the story and questions themselves. Second, instead of using the word *God,* I used the names of four different Hindu deities (Brahma, Shiva, Vishnu, and Krishna). As before, after completing the story comprehension portion of the task, I asked participants to complete a questionnaire concerning their stated ideas of the same god.

In fact I found little differences among the gods on the questionnaire or story comprehension task. What I did find was a large difference between what people said they believed about these gods and the concept of the gods that they used in making sense of the stories—just as in the United States.

Then came the surprise.

Another difference between the American studies and the Indian one was the age range of the participants. In the United States, people in their early twenties made up most of my sample. In India, I had participants ranging in age from nine years old up to fifty-five.

This forty-six-year age range allowed me to investigate whether the gap between stated beliefs about a god and the god concept used changed with age. To my surprise, I found a statistically reliable change—in the opposite direction of what I expected. The gap grew wider with age. Not surprisingly, adults tended to anthropomorphize God less than children in stated beliefs on the questionnaire, but they anthropomorphized more than children during the story comprehension task. Older participants made more intrusion errors suggesting that the gods have human-like limitations. These studies have recently been replicated in a much more rigorous and culturally sensitive manner by Travis Chilcott and Ray Paloutzian.[27] Again, older Hindus were more anthropomorphic in the story task than younger ones, independent of education or adult religious training.

More research on this topic would be helpful, but it looks as though children might actually be more nimble in their religious thought than adults in some cases—that the older we are, the less able we are to use ideas on the go that deviate too far from natural religion. This possibility converges with the idea of a sensitive period for religious development that I introduced in the previous chapter. Perhaps children lose natural facility in religious thought if they fail to adequately exercise it from an early age.

When I was a graduate student I had a friend who frequently referred to God as the "Big Guy in the sky." Sounds pretty childish, doesn't it? This cheekiness was not meant in any sacrilegious way, but indicated how comfortable this young man felt with God. He considered God a close friend as well as the ruler of the cosmos. This engineering student enjoyed reading theology and desired to go into full-time Christian ministry. It also became clear that he did not believe that God is actually a large-stature human being riding on the clouds. In fact, when asked, this young man claimed that God does not actually occupy space at all. Saying that God is here or there is misleading. God is not big, or a guy, or in the sky. Why then would my friend call God the "Big Guy in the sky"? In relaxed situations when he wanted to think about or talk about God, this very bright fellow simply found it easier to understand God as a big guy in the sky. He could more easily comprehend, relate to, and believe in a big guy in the sky than an omnipresent cosmic sovereign. No wonder, then, that Christians willingly sing that God "has the whole world in His *hands*" and that "His *eye* is on the sparrow" and even, "When He rolls up His sleeves he ain't just puttin' on the Ritz. . . . There is lightning in His footsteps and thunder in His fists."

SEVEN

It's Okay to Be Childish

Is BELIEF in God or gods infantile or childish? In certain respects, the answer is clearly yes. Belief in gods is likely to arise early in the development of a child. Before mastering riding a bicycle, knowing the boiling point of water or how to multiply, and even before learning how to read, children all over the globe already know about and believe in the supernatural beings talked about by their parents: ghosts, forest spirits, ancestor spirits, angels, devils, gods, *or* God. The tendency to believe in at least one superpowerful, superknowing, superperceiving, morally concerned God appears to be part of normal human child development before age five. Regarding a certain belief as infantile, however, bears little on whether we should continue to hold the belief in adulthood.

I presented some of the ideas in this book to an audience at Franklin and Marshall College in Lancaster, Pennsylvania, in autumn 2006. During the lively discussion session after the formal lecture, the audience seemed especially interested in the notion that the foundations of belief, or even belief itself, arise naturally during the preschool years. As I affirmed this interpretation of the evidence to date, a gentleman near the front asked pointedly, "Doesn't this mean that belief in God is childish?" The remarkably polite audience gasped in astonishment at what they regarded as a cheeky or even hostile question. But it was a fair question.

The question of whether belief in God (or any other superhuman agents) is childish, infantile, or some kind of residue of childhood naiveté arises repeatedly when the nature of belief is discussed. Isn't God the same as Santa Claus or the tooth fairy, a being that children believe in but then really should outgrow?

Attacks on Religion for Being Childish

One of the most famous versions of the religion-as-infantile argument was developed by Sigmund Freud. Multiple times in his book *The Future of an Illusion,* Freud voices his insistence that belief in gods arises as childhood anxiety projected onto the natural world:

> Now when the child grows up and finds that he is destined to remain a child for ever, and that he can never do without protection against unknown and mighty powers,

> he invests these with the traits of the father-figure; he creates for himself the gods, of whom he is afraid, whom he seeks to propitiate, and to whom he nevertheless entrusts the task of protecting him.[1]

According to Freud, adults retain the childlike fear of unknown forces and invent a god to serve as a cosmic father figure, reflecting the early childhood need to be protected by their father but simultaneously afraid of him.

Freud's analysis does not stop with any given individual child or any individual developmental course from childhood into adulthood. Rather, he is out to explain why it is that belief in gods is a human universal. To do so, he offers that the developmental story he promotes merely reenacts on the individual level a problem that humanity in general has faced from its origins as a species:

> And so a rich store of ideas is formed, born of the need to make tolerable the helplessness of man, and built out of the material offered by memories of the helplessness of his own childhood and the childhood of the human race.[2]

To sum, Freud sees belief in gods as infantile or childish to the core. We believe in gods because we carry around anxiety in the face of natural forces beyond our control. We cope with this anxiety by inventing a being or beings with the properties we attributed to our fathers during early childhood. We remember (if only unconsciously) feeling protected by a powerful but frightening father when we were infants. We use this infantile memory (again, unconsciously) to create a god concept. But

Freud goes one step further: belief in gods is an infantile *illusion* and an ancient one at that:

> Similarly man makes the forces of nature not simply in the image of men with whom he can associate as his equals—that would not do justice to the overpowering impression they make on him—but he gives them the characteristics of the father, makes them into gods, thereby following not only an infantile, but also, as I have tried to show, a phylogenetic prototype.[3]

By "phylogenetic" Freud means that gods are not only primitive in terms of individual human development but also a residue of our primitive, prehistoric ancestry. Crudely put, we get belief in gods from our cavemen ancestors, maybe even from prehuman species. Belief in god is infantile and prehuman.

Not many psychologists or other scientists of religion take seriously Freud's evidence-thin account of the origins of religious thought, but the idea that belief in gods is something embarrassingly childish has never completely gone away. To take an illustrative example, Richard Dawkins adopts a religion-as-infantile stance in some recent interviews and writings. Here is an illustrative example of Dawkins's perspective from an interview he gave in 2006:

> I do care passionately about what's true. One major difference between Santa Claus and God, obviously, is that no adult believes in Santa Claus, and unfortunately a great many adults believe in God. It's about time they grew up, and toss God aside at about the same age that they toss

> Santa Claus aside. If there are some people who are distressed by a loss of faith, I would encourage them to hang in there, because if you really stand up and look the real world squarely in the face, it does turn out to be a much more wonderful place than the sort of make-believe, childish world of religion.[4]

For Dawkins as well as Freud, belief in God or gods is childish and people should grow up and "toss God aside."

Labeling religious beliefs and practices as childish and thereby concluding that they should be abandoned on that basis amounts to nothing more than persuasive but empty rhetoric.

The common comparison with Santa Claus and the tooth fairy betrays disingenuousness, intellectual laziness, or serious ignorance. For starters, adults believe in God but not Santa and the tooth fairy, whereas young children may believe in all three. In fact, as scientist-theologian Alister McGrath has pointed out, many adults (himself included) come to belief in a particular god after having not believed as children. This fact too is a point of incongruity with the cases of Santa Claus and the tooth fairy. People do not begin or resume believing in them in adulthood after not believing in them as children.[5] Santa Claus and the tooth fairy also fail to fit the conceptual space that children (and adults) have because of their natural cognition. Santa Claus and the tooth fairy do not readily account for perceived order and purpose in the world, for great fortune and misfortune, for matters concerning morality, life, death, and the afterlife, and they have little relevance in day-to-day matters outside their very limited ranges of concern (Christmas presents and the tooth-for-money exchange). Note, too, that parents deliberately dupe children into such beliefs through stealth and deception. Adults do not (typically) eat the sacrifice placed out for gods and pretend that the gods ate it the way they eat Santa's cookies. If indoctrination and theatrical acts of deception were all gods had going for them conceptually, adults would outgrow them too.

I imagine most adults are sympathetic to the idea that "childish" or "immature" thinking should be abandoned in adulthood. It goes without saying, doesn't it? Please excuse my naiveté, but why? Why does an idea deemed childish automatically mean a bad, dangerous or wrong idea? You may be thinking, "Surely children know less than adults and make more mistakes in reasoning, and so their judgments are not as trustworthy." Agreed. What follows from this, however, is only that we should more carefully scrutinize the beliefs of children than those of adults, particularly if they deviate from what adults believe. But adults generally *do* believe in gods.

That belief in gods begins in childhood and typically continues into adulthood places it in the same class as believing in gravity, the permanence of solid objects, the continuity of time, the predictability of natural laws, that causes precede effects, that animals bear young similar to themselves, that people have thoughts and wants that motivate and guide their actions, that some things are morally right or wrong, that their mothers love them, and numerous other ideas about the world, some of which have been discussed earlier in this book. These beliefs all arise early in childhood and typically persist into adulthood. If believing in gods is being "childish" or "immature" in the same respect as these sorts of beliefs, then belief in gods is in good company.

I favor the approach that regards our minds as basically trustworthy to deliver true beliefs and that our naturally arising "childish" beliefs should be regarded as true until we have good reason to suspect them as being problematic. It is not clear to me that we can do otherwise and still function as normal, sane human beings. So much of our core knowledge and guiding values arise during childhood and shape our lives. We should trust these "childish" beliefs as innocent until proven guilty. A Darwinian atheist may reply that there are other good reasons to the contrary of belief in many, if not all, gods. Perhaps so. But the discovery that theism is "childish" is not one of those good reasons.

Do bear in mind too that just because an idea or belief does not typically arise in childhood but waits until adulthood does not make the belief true. If you consider belief in gods to be "immature" or "childish" because of its early developmental history during the life span, then keep in mind that many "mature" or "adult" beliefs bear no advantages over childhood beliefs in terms of truth, value, or desirability. Adults come up with scientific theories and philosophical positions that we later discard as false or unhelpful. A child never believes that nothing really exists except the self or that there is no external world or that we are brains in a vat somewhere. Adults try on these kinds of beliefs. Adults thought that tobacco smoking and consumption of narcotics are harmless and enjoyable pastimes and that bleeding people or cutting holes in their skulls are good treatment for mood disorders. Children do not come up with stuff like that. Adults are more inclined to believe that killing themselves or someone else is a good idea. You would be hard pressed to find a five-year-old agreeing with you under any conditions. Adults may be more knowledgeable than children and perhaps cleverer, but in addition to being more right, this means adults can be more wrong.

By raising these issues, I mean only to show that whether a belief is "childish" in the sense of emerging in early childhood has no immediate bearing on whether it is true or good. It may be that it is most prudent to treat childhood beliefs as innocent until proven guilty.

Virtuous Childishness?

Is belief in God or gods infantile or childish? In certain respects the answer is clearly yes. Belief in gods is likely to arise early in the development of a child. As I hope I have demonstrated thus far, the cognitive tools needed to get belief in gods off the ground are in place during the preschool years.

Regarding a belief as infantile, however, bears little on whether we should continue to hold the belief in adulthood. Once the rhetorical name-calling is stripped away, one can see that having foundations in childhood is irrelevant to whether a belief is true or false. Calling a belief infantile or childish amounts to nothing more than name-calling, a cheap trick for trying to scare someone away from the belief.

Jesus appeared to have little fear of this guilt by association with children. Rather, when his followers tried to shoo away children, he reportedly replied by saying, "Let the children come to me, and do not hinder them, for the kingdom of heaven belongs to such as these."[6] He regarded something about childishness as a welcome, and not shameful, attribute in his followers. Exactly what Jesus regarded as the desirable trait or traits that children possess is an open theological question. Humility is a likely part of the answer, for in another episode Jesus explained, "I tell you the truth, unless you change and become like little children, you will never enter the kingdom of heaven. Therefore, whoever humbles himself like this child is the greatest in the kingdom of heaven."[7] Maybe, too, author G. K. Chesterton was on the right track when he wrote that children possess an "unspoilt realism" that enables them to see things as they really are—to see the good, bad, puzzling, disturbing, or miraculous for what it is instead of adding adult-like layers of rationalization and self-delusion or fancy theoretical gloss that makes things appear other than they are. Instead, Chesterton suggests that to properly understand and believe about God, adults need to "invoke the most wild and soaring sort of imagination; the imagination that can see what is there."[8] Or perhaps Jesus referred to the way in which children seem able to receive positive regard, gifts, and even forgiveness without feeling embarrassed or indebted to the benefactor. Could this be part of the humility he valued? For adults, grace and graciousness may go against the grain of our embedded senses of proper social exchange or of a just world (as I developed in the previous chapter). Perhaps Jesus meant to emphasize children's ability to trust unswervingly, not needing to understand all of the hows or whys but only needing to understand the who. Whatever Jesus was driving at in saying that his followers should become like little children, he refused to be shamed by accusations of his followers as being childish in their belief.

EIGHT

So Stupid They'll Believe Anything?

> I am completely unreligious. It is so strange that my 4 year old believes in god and talks about it once in a while. I never taught that to him. Anyway, sounds interesting, it's partly human nature.

This quote, a blogged note in response to a *London Times* online article about one of my born believers lectures, speaks directly against the idea that children believe simply because, and only because, their parents indoctrinate them.

Simply put, the indoctrination hypothesis is the claim that children come to believe in the gods—any and all—because they are thoroughly indoctrinated in those beliefs. From early in life they hear about Allah, the ancestors, Jesus, or Vishnu.

Parents and other authority figures explicitly teach children all about these gods' many properties. Further, parents impress on children through terrible stories about eternal infernos, famine, storms, and volcanoes that failure to believe in and obey these gods carries severe consequences. Indeed, some parents and religious teachers physically punish children who express doubt or deviate from the god-mandated taboos and directions. And when children express belief and obedience to the gods, they are praised, rewarded, and encouraged with promises of idyllic afterlives or even riches and success in this world. Under such conditions, children would believe anything that they cannot otherwise disprove, and gods are notoriously hard to disprove or prove. This fully explicit, coercive indoctrination system is why children believe in gods of one sort or another, according to the indoctrination hypothesis.

Maybe people who suggest the indoctrination hypothesis have in mind cases such as Jim Jones's People's Temple tragedy or David Koresh's Branch Davidians—cases in which seemingly ordinary, sensible adults adopted radical new religious beliefs, abandoned their past careers and families, and, in many cases, forfeited their own lives for their beliefs. If adults can succumb to the pressures of systematic indoctrination, what chance do mere children have against its influence?

In this chapter I discuss the indoctrination hypothesis and some of its near relatives because of its importance for those who wish to teach children to believe (or not to believe) various religious ideas. Can children be successfully brainwashed into believing just about anything?

Philosopher Anthony Grayling expressed the notion this way in reference to Christian higher education:[1]

> Remember that all this Christian teacher training is aimed at religious brainwashing of the young, not least the very young. Without brainwashing of the young, religion would wither and die of its own absurdity. The religions—all of them—depend crucially on recruitment by capturing the minds of children.

But would any and all religion "wither and die of its own absurdity" without "brainwashing" children? I am afraid there is no good evidence supporting Grayling's speculation and plenty of evidence to the contrary.

As initially sensible sounding as the indoctrination hypothesis may be, and though I have had it suggested to me numerous times in informal conversations or in discussion sessions after academic lectures, this hypothesis has received little attention from scholars of religion. Several reasons for this neglect spring to mind.

Foremost, cultural anthropologists, religious studies scholars, and people raised in religious communities find the indoctrination hypothesis a caricature of what typically happens in religious communities. Rather than coerce, threaten, and bully children into belief, adults simply believe in the forest spirits, ancestors, witches, or God and act accordingly. They conduct the appropriate rituals, say prayers, discuss the meaning of life events, wonder about the activities of gods, and go about life as if gods were just as natural, normal, and certain as air, gravity, or germs. Ethnographies of religious belief and practice in traditional societies often stress the commonness of religious discourse and how it is neatly woven into daily life.[2] Not until the past couple of hundred years, and only still in a minority

Adding to testimony a healthy dose of intimidation does not necessarily make things any better for the would-be indoctrinator. Outside of pre–Geneva Convention military-prison-camp-type brainwashing techniques, such strategies may backfire. Threatening dire consequences of disbelief or otherwise making an idea, such as the nonexistence of God, emotionally charged may mark out the idea as unusual or special, and thereby increase the likelihood of children entertaining it. We simply have no reliable scientific evidence that this kind of indoctrination strategy works well in the course of child rearing.

Another new atheist, Christopher Hitchens, partially acknowledges the weakness of the indoctrination hypothesis while maintaining its importance:

> Indoctrination of the young often has the reverse effect, as we also know from the fate of many secular ideologies, but the religious will run this risk in order to imprint the average boy or girl with enough propaganda. What else can they hope to do? If religious instruction were not allowed until the child had attained the age of reason, we would be living in quite a different world.[3]

So indoctrination sometimes backfires, but the religious have no choice but to rely on it, because without it, no one would become religious. By now you already know that the evidence sides against the idea that all religion has going for it is indoctrination, but allow me another illustration of the point relevant to Hitchens's claim.

Suppose indoctrination really was the only tool available to "the religious" but it sometimes has "the reverse effect," and so, say, one out of every ten children becomes a nonbeliever even with indoctrination. Let us suppose, as Hitchens and Grayling seem to suppose, that people do not have any natural dispositions toward religiousness, and so if they were not religiously indoctrinated, or were equally taught religious and nonreligious ideas, the nonreligiousness would win out. By inability to retain their numbers, religious people would be less than one-twentieth of the population being religious after six generations.[4] If childhood indoctrination were the lifeblood of religion, Christianity, Hinduism, Judaism, and Islam would not last a couple of centuries. Yet here they are. Something else has got to be going on.

The indoctrination hypothesis also suffers from the same shortcoming as other passive enculturation accounts of why people believe what they do; the problem is simply pushed back a generation. Why do people believe in gods? Because their parents and other elders indoctrinated or (more charitably) enculturated them. But why did their parents and elders believe in gods? Because their parents and elders enculturated them. And so forth, and so on. I suppose under this sort of account the reason belief in ghosts is so widespread goes something like this:

> Once upon a time, by random chance, someone decided to believe in ghosts. (It could have been invisible potatoes, transdimensional cows, flying spaghetti monsters, mind-reading socks, or something altogether different, but it happened to be ghosts.) This ghost believer then indoctri-

> nated his children to believe in ghosts, who indoctrinated their children to believe in ghosts, and so on. With each generation, it got easier to indoctrinate. Belief became so widespread it might be considered a cultural belief. With so many believers around, indoctrination became enculturation. More and more people in the community believed in ghosts, so more and more people could help with the enculturation process. That is why so many people believe in ghosts.

Some people (academics included) seem to find such explanations satisfying. I am not one of those people. I want to know why *these* beliefs and practices instead of others, and why these beliefs spread so successfully to other people whereas others presumably did not. Why not mind-reading socks? Why not transdimensional cows? Simple indoctrination or enculturation offers no such explanation.

Finally, Grayling and Hitchens's claims that religions must indoctrinate young children or else they would disappear has an ethnocentric ring to it. Many religions around the world systematically exclude children from their central rites and beliefs. For instance, anthropologist Frederick Barth's famous study of the Baktaman people of New Guinea carefully describes a set of practices and beliefs that are exclusively the domain of adult males. As males reach certain levels of maturity, they are initiated into the secrets of the tradition in a stepwise fashion. Older men know more than younger men, who know more than adolescent males. Women and children are formally excluded. As so many ethnologists have shown, such a pattern is not unusual in traditional societies' religious practices. These cases severely undercut the claim that brainwashing the young is critical for a religion's survival. The democratic, nearly full inclusion of children (and women) in the beliefs and practices of Christianity may actually be a cultural and historical anomaly.

The one piece of "evidence" I hear most recently cited in relation to the indoctrination hypothesis is the observation that children tend to "inherit" their religion from their parents. Hindu parents have children who grow up to have Hindu beliefs. Muslim parents have children who grow up to have Muslim beliefs. On a more finely grained level, Lutheran, Baptist, and Methodist parents tend to have Lutheran, Baptist, and Methodist children, respectively. The observation that children have a tendency to adopt the religious beliefs of their parents supports the contention that children are influenced by their parents and the communities in which they are raised. If a child did not grow up with Hindu parents and was in no contact with other Hindus, it would be unlikely that the child would grow up to be Hindu. It does not follow, however, that if a child grew up with no Hindu parents and in no contact with other Hindus, the child would likely grow up without any religious beliefs and practices at all. As the indoctrination hypothesis attempts to account for why people have any and every kind of religious belief, the observation that specific varieties of religious beliefs are influenced by parents and social environment is irrelevant.[5]

Compare the "inheritance of religion from parents" to the inheritance of food preferences from parents. Surely people who grow up in a family where breakfast cereal, eggs, and toast are customary will be more likely to grow up eating breakfast cereal, eggs, and toast rather than refried beans, fruit, and sweet bread. Nevertheless, we cannot conclude from this fact that there are not natural predilections to eat some things over others. We would be silly to say that indoctrination at the hands of parents explains why people generally prefer sweet foods over bitter ones or why people tend to eat more grains (rice, wheat) than arthropods (ants, spiders, beetles). Food preferences have a large natural component that gets tuned up and specified by living in particular cultural environments. Similarly, religious belief and practice is largely natural but gets tuned up and specified.

Evolved Gullibility?

For the reasons above, I am unmoved by a recent reformulation of the indoctrination hypothesis offered by biologist Richard Dawkins in *The God Delusion.*[6] In fairness to Dawkins, more than once he notes that his argument for why children come to believe in gods is meant only to be illustrative and perhaps only complements others' accounts. Nevertheless, because some people have sympathies for such stories and because Dawkins is such an effective communicator, I fear some readers may have taken his offering more seriously than Dawkins himself does.

Beyond the explanations he favorably reviews from others, Dawkins makes two contributions of his own. The first is a version of the indoctrination hypothesis we might call the *evolved gullibility hypothesis*. Second, Dawkins amplifies this evolved gullibility hypothesis with his famous meme theory, according to which cultural concepts, by analogy with genes, replicate better or worse because of their survival traits. Dawkins's suggestion, the evolved gullibility hypothesis, is that humans have evolved such that their children are gullible recipients of whatever their parents tell them. If parents tell them about a god, they believe it, wholly and unquestioningly. Dawkins has given evolutionary reasons for why indoctrination is successful. Add to this gullibility the claim that belief in gods is particularly successful at spreading—a "mind virus," or meme—because of the alleged grave consequences of not believing (or even questioning the idea of gods), and you have the basis for why belief in gods persists. Dawkins's stature as a famous popularizer of science makes his claims in this instance worthy of closer scrutiny, especially as similar themes appear in other recent anti-religion books by Daniel Dennett and Christopher Hitchens.[7] His claims also happen to be far from plausible. Dawkins sums up his twofold suggestion this way:

> If I have done my softening-up work well, you will already have completed my argument about child brains and religion. Natural selection builds child brains with a tendency to believe whatever their parents and tribal elders tell them. Such trusting obedience is valuable for survival: the analogue of steering by the moon for a moth. But the flip side of trusting obedience is slavish gullibility. The inevitable by-product is vulnerability to infection by mind viruses.[8]

Children have evolved gullible brains making them susceptible to "mind viruses." In this context, religious ideas are the "mind viruses" that infect defenseless young brains.

The reason children have evolved gullibility is our species' need to successfully pass on cumulated knowledge, particularly to children who do not reach self-sufficient maturity for a long time in comparison to other species. Dawkins writes, "More than any other species, we survive by the accumulated experience of previous generations, and that experience needs to be passed on to children for their protection and well-being."[9] He explains:

> For excellent reasons related to Darwinian survival, child brains need to trust parents, and elders whom parents tell them to trust. An automatic consequence is that the truster has no way of distinguishing good advice from bad. The child cannot know that "Don't paddle in the crocodile-infested Limpopo" is good advice but "You must sacrifice a goat at the time of the full moon, otherwise the rains will fail" is at best a waste of time and goats. Both admonitions sound equally trustworthy. Both come from a respected source and are delivered with a solemn earnestness that commands respect and demands obedience. The same goes for propositions about the world, about the cosmos, about morality and about human nature. And, very likely, when the child grows up and has children of her own, she will naturally pass the whole lot on to her own children—nonsense as well as sense—using the same infectious gravitas of manner.[10]

From this perspective, children are profoundly gullible regarding some of life's most weighty matters. Is this claim accurate?[11]

As someone who has spent a fair amount of time around children, I have some sympathy with the idea that children are strikingly gullible. Even teenagers can fall prey to a susceptibility to believe those in authority. Young Life, a youth organization I once worked for, aggressively guards surprise in its camping programs, and so when teens ask, "What are we going to do next?" leaders often resort to tongue-in-cheek deception (especially after they are exhausted by saying, "Just wait and see," or "It's a surprise," or "I'm not going to tell you"). So I have heard leaders say things like, "We're going to count all of the gravel in the parking lot." About one-fourth of teenagers believes it (and registers complaint at the prospect). Tell them not to walk on the grass because it is a designated "nematode [roundworm] sanctuary," and they buy it. I knew a thirteen-year-old who thought that a particular spray breath freshener marketed as "instant Irish accent" could really give him an Irish accent!

It is not hard to find examples of what looks like broad gullibility in children, teens, and even adults (think of some of those tabloid headlines). One of the highest teaching priorities for high school and university teachers is to help their charges overcome a tendency to accept ideas without critical scrutiny. The top-selling introductory psychology text for university students that I have used for my classes includes an early chapter on healthy skepticism, a foundational principle for becoming a scientist, but one hard to learn.[12] The outrageous claims blaring from newsstand tabloids and the rampant spread of urban legends on the Internet testify to how susceptible adults are to evidentially shaky claims. If adults are gullible, certainly children are too. Right?

It also sounds sensible that there might be a survival advantage for those who rapidly believe the authoritative testimony of parents and others. If you spend all of your time demanding corroborating evidence or a second opinion, you will never learn anything. Reserving judgment until all the facts are in could be fatal. When the shepherd boy calls "Wolf!" we better give him the benefit of the doubt, at least the first or second time, or our children and livestock could suffer from our skepticism.

As with many other initially plausible-sounding claims, however, closer scrutiny reveals numerous limitations and problems for the evolved gullibility hypothesis. In short, evidence from developmental psychologists suggests that though children (and adults) do have a tendency toward credulity, this willingness to believe is not uniform with regard to all kinds of ideas or all kinds of sources. Rather, we are biased to believe some sorts of things over others, from some people over others.

Dawkins exclaims that children swallow whole parents' or parent-ordained authorities' "propositions about the world, about the cosmos, about morality and about human nature."[13] But the scientist Dawkins cites also claims in that very book that children's acquisition of beliefs is naturally and fundamentally biased in some directions over others in the domains of the world, cosmos, morality, and human nature. Children will not simply believe any and all propositions with equal zeal.

For instance, one straightforward implication of Marc Hauser's book *Moral Minds* on the evolution of moral sensibilities, which Dawkins so warmly cites and appropriates in the very next chapter after his discussion of evolved gullibility, is that you cannot simply teach children just anything about morality.[14] According to Hauser, they possess a moral instinct (analogous to a grammar) that informs and restricts the range of moral rules that are likely to be received and understood as unchangeable norms. Hauser's is one explanation for why it is that basic moral norms—do not murder members of your own group, steal possessions, or steal spouses from members of your group, needlessly harm others in your group, and so forth—recur across cultures. Moral intuitions are not arbitrary and do not successfully spread only because of evolved gullibility.

The same curious oversight extends to Dawkins's evolved gullibility hypothesis as applied to "the world," "the cosmos," and "human nature." Research by the very same scientists, in the very same publications that Dawkins cites approvingly in the same chapter of his book where he makes this claim, have demonstrated that the evolved gullibility hypothesis cannot blandly extend to "the world," "the cosmos," or "human nature." Indeed, Scott Atran, Pascal Boyer, and Deborah Kelemen argue to the contrary that children are biased to adopt some beliefs over others by the way their minds naturally function. Atran puts the place of belief in gods in startlingly strong terms when he writes: "Supernatural agency is the most culturally recurrent, cognitively relevant, and evolutionarily compelling concept in religion. The concept of the supernatural is culturally derived from an innate cognitive schema."[15] His point is that our biology biases us toward belief in gods. Blind gullibility does not and cannot do the job on its own. Similarly, in Chapter 2, I shared some of Kelemen's research that demonstrates that young children are biased to see the natural world as purposefully designed by some kind of agent. It seems that propositions to the contrary, such as natural selection accounting for the apparent design in living things, find in children a far less receptive audience. Let a trusted authority figure tell six-year-old children that animals evolved from different animals and have another trusted authority figure tell the same children that animals were created purposefully by a god, and the science to date suggests that the majority of the children will believe in the creation account.[16] Children's minds are simply more receptive to some ideas over others (adults too, for that matter).

Kids Have to Understand or They Can't Believe

One relatively straightforward way in which at least a strong version of the evolved gullibility hypothesis can be misleading is the fact that it is harder to believe something you do not understand than something you do understand. When I homeschooled my daughter and I tried to teach her mathematics, sometimes she just could not understand what it was I was trying to teach. It was as if my words simply bounced off her head without penetrating at all. The problem was not that she was not inclined to believe me. Rather, some of the particular beliefs I was trying to pass on made no sense to her. If I tell a child that most of the stuff in the universe is invisible dark matter that cannot be seen but can exert gravitational force on stars and galaxies and so forth, the child might nod along and apparently believe me. If what I am saying sounds to the child the same as "Most of the stuff in the blah blah is blah blah that cannot be seen but blah blah blah on stars and blah blah blah," however, it is a stretch to say that the child has now acquired a new idea that she or he believes. Similarly, if I try to convince my child to believe that there are special potatoes that exist in seven dimensions of space and time, can read minds of people before they have thoughts, vanish anytime anyone looks at them, and pass in and out of existence every prime-numbered day of the month that falls on a Wednesday, but matter terribly to whether each person's ether passes into the prior world after reincarnation as a protozoan, the child (or adult) is probably going to reply something like, "Hmmm, potatoes. I like them fried." Some minimal understanding on the part of a child is necessary for her to acquire a belief, and her conceptual abilities restrict what it is she understands. Hence, what children believe is restricted by their conceptual abilities, no matter how enthusiastically or forcefully endorsed an idea is by parents. When it comes to religious beliefs, if the ideas are too counterintuitive and thereby fall too far outside children's conceptual abilities, they will not be believed. On the other hand, concepts that are readily conceived of *and* fit with their naturally arising conceptual biases will be more likely to be believed.

In a related way, depending on the age of the child and the type of information, children may simply be unable to use a parent's testimony. Recall John Flavell's experiment with the blue and white cup described in Chapter 5. The adult on the other side of the barrier from the child and the blue cup says clearly, "I can't see the cup. Hmm. I think you have a white cup over there. I think you have a cup that is white." The adult (Ellie) just told the child about her belief—the kind of unobservable thing that children must rely on the testimony of others to learn about. But we already have seen that many three-year-old children simply cannot use this testimony. Contrary to what Ellie has said, they insist she believes the cup is blue.

The evolved gullibility hypothesis is not out of the woods when children are four or five either. By this age, many children can successfully use Ellie's testimony and are prone to give it the benefit of the doubt a lot of the time, just as adults do. The children's understanding of minds has developed enough that they understand that Ellie could have a mistaken belief, enabling them to believe her statement about her belief. Notice, however, that this understanding about the potential fallibility of belief simultaneously enables them to be skeptical about what others say. For instance, if a five-year-old child knows that you were out of the room when the chocolate was moved from one cabinet to another, she knows that you are not a trustworthy source for telling where the chocolate is. Your testimony, in this regard, is not taken very seriously. (At this age, too, children begin trying out genuine lying for themselves.) So before children understand that beliefs can be wrong, they simply cannot use all adult testimony. After they understand that beliefs can be wrong, children know not to use all adult testimony. A simple, strong version of the evolved gullibility hypothesis will not do.

Some Sources of Testimony Are Better Than Others

Another complication with a strong version of the evolved gullibility hypothesis is research indicating that people (from childhood) are very sensitive to who constitutes a good model for imitation. As anthropologist and psychologist Joe Henrich has demonstrated, perhaps for reasons of natural selection, people do not just blindly follow the example (spoken or otherwise) of any parent or adult in authority.[17] Rather, they identify role models most appropriate for the kind of information or skill that they want to learn. For this reason, boys in traditional societies tend to imitate high-status, successful males regarding practices they must master, such as hunting, forest navigation, and warfare. What their own mother says about these aspects of the world and human nature is less important than what the village war hero or top hunter has to say. This selective attention may be especially important during the teen years. If my wife gives my son advice concerning basketball, he turns to me and says, "What do you think, Dad?" If I volunteer fashion insights to my daughter, she rolls her eyes and consults her mother. Gullibility is limited by the source. This form of "gullibility" is not so different from what even educated and healthily skeptical adults practice. If an apparent expert speaks about something we care about, we listen.

Inadvertently (I hope) Dawkins has suggested a view of the developing human child's mind that assumes the young mind is a blank slate just waiting to be filled in. As Nicholas Humphrey has said, "Children are made of the words they hear."[18] Such a perspective lurks in many social sciences but has not squared with the state of the art in the psychological sciences for at least three decades.

When my daughter was seven years old, my father and I took her and my son on a canoeing trip in a wilderness area. Walking back from the boats, we came across a snake sunning in the middle of the path. The children's beloved and esteemed grandfather observed that there are really only two types of snakes in the world: rattlesnakes and cobras. As this snake before us was clearly not a rattlesnake—not possessing rattles—it had to be a cobra. Knowing that cobras were very dangerous snakes, how did the seven-year-old girl receive this information in this context that clearly would have exerted selective pressure on our evolutionary ancestors? She picked up the snake. Childhood gullibility is not as simple as it might at first appear.

Again, in fairness to Dawkins, his proposal about how childhood gullibility in conjunction with the sorts of biases identified by Atran, Boyer, Evans, Hauser, Kelemen, and others may require only a fairly weak version of the evolved gullibility hypothesis. The natural early-developing features of human minds bias children (and adults) toward belief in gods and to adopt particular moral attitudes and then a general (but not absolute) tendency to believe whatever they are taught entrenches particular variants on those religious beliefs. Natural mental architecture anchors the religious ideas, but receptivity to adult testimony helps spread, amplify, and even diversify these religious beliefs. If this is what Dawkins is getting at, my position is not so very different (at least on this matter). Dawkins does, however, develop this story differently than I would. In particular, I have reservations about comparing religious beliefs to viruses of the mind.[19]

GOD IS MORE LIKE *E. COLI* THAN INFLUENZA

The religion-as-virus idea, or the infectious meme hypothesis, has scarcely gained any mention by the scientists who actively research the natural foundations of religion. It simply is not an explanatory framework that has proven itself valuable in the cognitive science of religion. Nevertheless, popularizing new atheists such as Dawkins and Dennett prominently feature versions of the infectious meme hypothesis in their accounts. Why this disconnection with the scientific literature? The lead possibility is that the metaphor of religion as a dangerous mind virus is catchy and plays into the antireligious agenda of these authors. They want to communicate the following line of reasoning. Viruses, like influenza, are bad, causing suffering and death. We try to kill viruses that infect us. Religion is a virus. Therefore, religion is bad and we should try to kill it. Indeed, the space in Dawkins's and Dennett's books devoted to showing that "religion" is bad outstrips the evidence showing that the analogy with viruses is scientifically helpful.

What I hope is clear from what I have presented in this book is that those of us who study religious thought and action—scientists and scholars such as Scott Atran, Jesse Bering, Pascal Boyer, Stewart Guthrie, Brian Malley, Bob McCauley, Deborah Kelemen, Tom Lawson, Ilkka Pyysiäinen, Jason Slone, Richard Sosis, Todd Tremlin, Harvey Whitehouse, and David Sloan Wilson—do not see religious ideas as intruders into human nature but as a wholly expected extension of the way humans are naturally put together. Belief in gods is not an invading parasite or an influenza virus against which we can be vaccinated or that can be removed from humanity.

Discovering the naturalness of religion is more like discovering that within our intestines live countless billions of the bacteria *E. coli.* When contaminating our food or drink, *E. coli* can cause illness or death. When escaping into our abdominal cavity, it can cause lethal peritonitis. In our urinary tract, it can cause severe infection, and in our cerebral spinal fluid, it can cause meningitis. *E. coli* is a killer. And sometimes religious expression causes suffering, pain, and death. The solution for *E. coli* and religion? Cast them out! Kill them! Rid these plagues from humanity!

Not so fast. Like *E. coli*, religious expression is an inextricable part of human life. *E. coli* does lead to illness and death under certain rare and largely preventable conditions. In its proper place and proper function (from a human perspective), however, *E. coli* is essential for normal human digestion. *E. coli* living abundantly in our large intestine helps us digest foods and produces nutrients that we require to live. If we killed all the *E. coli* living inside us, we could end up killing ourselves in the process. So, too, with religion. Religious thought and devotion can lead to pain and suffering in death under peculiar, largely preventable conditions. Normally, however, religion is a fundamental and healthy part of human existence, springing from cognitive systems that if removed would remove our humanity. Religious thought and action is so integral an expression of normal human nature that the cure for it could kill the patient.

NINE

Is Atheism Unnatural?

When I have delivered lectures and explained how the scientific study of religion points to the naturalness of religious belief, and particularly how well tuned children are to believe in God, about one out of every fifty university audience members asks something along the lines of, "If religion is so natural, then what's wrong with me?" After a recent lecture at the University of Copenhagen, an audience member commented that she could not remember ever having believed in God. What accounts for this unbelief? Perhaps throughout this book you have had similar questions nagging at you. If religion is so natural, then does that mean atheism is unnatural and requiring a special explanation?

This is an area needing much more scholarly and scientific

attention, and what I have to offer amounts more to a way to make sense of general observations than a fully developed and scientifically tested theory. Some people become atheists—not just uninterested in God or other supernatural beings but dismissing the existence of any supernatural agents, including God, spirits, ghosts, angels, devils, witches, and fairies. I will not be offering reasons in support of atheism or a defense of atheism, but I am suggesting the sorts of factors that might contribute to someone being resistant to religious thought.

Atheism Is Rarer Than You Might Think

If you are one of those people who never recalls having believed in any kind of god—including ghosts, fairies, angels, and the like—then the first thing you must understand is that you are very unusual. That you are special in this regard may be hard to believe. You probably know of at least a handful of other people who, like you, believe that they have never believed. But do not fall victim to what psychologists call the *availability heuristic,* a tendency to form a judgment of how common something is based on how easy it is to think of examples.[1] Such a reasoning strategy works fine a lot of the time but also leads to gross misestimations. The availability heuristic leads people to regularly overestimate how representative they and their social group are of people generally. Sadly, we fall easy prey to this egocentrism even when we know better.

Also keep in mind that we tend to interact most with people like us. Research by social psychologists has demonstrated time and again that we like people who are like us and "birds of a feather flock together."[2] Unless atheists are different in this regard, atheists are more likely to hang around with other atheists, further amplifying possible error in using the availability heuristic.

Atheist academics working in university settings may be especially prone to overestimate how common bona-fide atheism is. For various social and political reasons, atheists happily share that they are atheists in public settings, whereas believers in gods, ghosts, and spirits are less likely to share their commitments. At a recent conference (about religion), I noticed that more than half of the nonbelieving presenters managed to proclaim their atheism in their presentation, whereas I did not hear a single theist (and there were many) mention his or her own religious beliefs during their presentations (with the possible exception of one speaker who seemed to suggest a belief in God and atheism at the same time). Atheists who only hear their colleagues affirm atheism are even more likely to think that everyone around them is an atheist.

So how many people are really atheists? It is hard to know. If anything, surveys of contemporary beliefs—such as those that suggest that only one in twenty Americans does not believe in the existence of God—probably underestimate the number of people who believe in supernatural beings. Also, these surveys fail to account for people who might not believe in "God" (usually understood as the Jewish or Christian deity) but believe in ghosts, spirits, Gaia, the Oversoul, Absolute Consciousness, or any number of other gods (under my cognitive science–inspired definition). A member of my family falls in this category. If you asked, she would tell you she is an atheist but she believes in ghosts. Another friend of mine disavows God but believes that the dead can communicate with the living. Yet another friend claims not to believe in God because she is angry at God. (I am not sure that you can be mad at someone *and* disbelieve in them.) Often measuring who is and is not a supernaturalist in the cognitive sense is a little like trying to measure who is a racist—some people will deny it, but their behaviors tell a different story. For instance, recent research in China reveals that though less than 10 percent of Chinese consider themselves Buddhist, about half have prayed to a Buddha or bodhisattva in the past year.[3] Even with these measurement challenges, Europe may be the only continent where nonbelieving adults approach a third of the population (some estimates are much lower). In Africa, North America, South America, and most of Asia, adult atheists are hard to find. Less than a hundred years ago and as far back before that as we can tell, essentially everyone believed in some kind of god. So, taking human existence as a whole, throughout the span of our species, belief in gods is the norm and nonbelief has been very unusual indeed.[4]

People who do not believe in any gods (broadly construed) are fairly rare in the grand scheme of things, and people who have never believed in any sort of superhuman agent are rarer still. What then about the woman who said she had never believed in God? I certainly do not deny that there are such individuals, but if you think you might be one of them, be careful not to jump to conclusions. Not remembering having ever believed in any gods, ghosts, and the like is different from not actually having ever believed. People are notoriously bad at accurately remembering early childhood, a phenomenon termed *childhood amnesia.* Things we think we remember—even with great vividness and confidence—are often false memories.[5] I would wager that a large proportion of people who think they never had beliefs in the supernatural in fact did and have since forgotten.

Those preliminary observations aside, people who do not believe do exist. Is their lack of belief so unnatural? If "unnatural" means sick, demented, or psychologically disturbed, then no. If "natural" means a typical expression of human nature, then, yes, their lack of belief is unnatural. If unnatural means not well supported by ordinary maturationally natural cognitive systems, then, yes, atheism is unnatural—but then so is being a concert pianist, a top-rate scientist, or a contemporary theologian.

This lack of belief in supernatural agents when believing them is a common human tendency that might be accounted for by a combination of personal and social or environmental factors. By personal factors, I mean the combination of biological and psychological factors that vary across individuals. By social or environmental factors, I refer to those conditions around us—where we live, whom we interact with, and so on—that influence our beliefs and behaviors. I take these factors to have a causal impact on which ideas are likely to be entertained, which will become beliefs, and which will spread broadly across populations. I distinguish factors from reasons someone might give for belief. Someone may have perfectly good reasons to justify his beliefs and also be subject to factors that have contributed to the likelihood that he believes one thing over another. Below I describe some factors that may contribute to the likelihood of being an atheist but leave aside reasons.

Personal Factors

For any given capability that arises naturally in a large population, there will be variability. Take walking, a great example of a natural ability. Though the vast majority of adults walk very naturally and normally, there is some variability. Some people develop with legs of slightly different lengths, making comfortable walking more difficult. Some people have weaker knee joints, reducing their walking endurance. Some people have problems in their brain's motor systems for coordinating movement, making walking less smooth and sure. We have all seen people who walk with some difficulty or require the aid of a cane or prosthetic to help them walk. In the extreme cases, such as people born without fully formed legs or with severe nervous system disorders, walking may prove impossible. That such people occasionally exist in a population does not mean that walking somehow is not natural.

Not believing in any sort of gods may prove to be a trait that is analogous to not being able to walk. Some people might have personal factors such as a biological and psychological endowment that makes such belief extremely difficult—perhaps requiring heroic intervention or "prosthetics" to make supernatural belief possible. The scientific study of atheism has only just begun, and so I can offer only speculative suggestions as to what sort of personal factors might prevent someone from naturally becoming a believer.

One factor that numerous cognitive scientists of religion have emphasized as central to normal religious thought, action, and experience is our theory of mind system and related social reasoning. In short, without a robust ability to reason about the mental states of other minds, it is particularly hard to reason about invisible nonhuman minds such as the minds of ghosts, spirits, and gods. As with any maturationally natural capacities, how readily and fluently theory of mind functions varies within a population. Psychologist Simon Baron-Cohen has developed a line of research investigating this sort of variability in theory of mind and related social cognition. He has controversially labeled the absence of strong social cognitive abilities and interests *male-brainedness.*[6]

He calls this personal endowment *male-brainedness* instead of *maleness* because it occurs in some women as well as men and varies greatly in men from very strong to essentially absent. Having a Y chromosome does not appear to be what is critical for being male-brained, though men are many times more likely to suffer from severe male-brainedness. A biological factor that does strongly predict male-brainedness (already showing signs in the crib) is the level of testosterone in the womb.[7] In its clinical extreme form, male-brainedness characterizes autism, argues Baron-Cohen.

Within hours of birth, the strongly male-brained person already has a preference for looking at mobiles instead of human faces. During the toddler years, the male-brained person prefers to play with blocks and toy machines instead of playing with dolls and playing house. By the school years and into adulthood, the male-brained person shows strong skills in systemizing, showing great interest and ability in making sense of how things work, in seeing order in complex systems. They may be drawn to the sciences and mathematics or engineering and building. Or they might apply their systemizing interest to human systems such as political and economic systems, but they are not generally what you would call "people persons." As strong as male-brained people are in systemizing, they are weak in empathizing. They generally show weaker appreciation of others' mental and emotional states and might be largely unaware of this relative weakness. The male-brained person does not fluently keep track of others' beliefs, desires, and motivations. A male-brained person is likely to be much more comfortable reading *Popular Mechanics* and popular science writing than *People* magazine and Jane Austen novels. Note that being somewhat male-brained is not pathological and does not necessarily lead to any noticeable living problems. In fact, male-brained people might be among the high achievers in many areas of public life, including in the academy.

If theory of mind and related social cognition are so critical for theistic belief and if severe male-brained people are weak in or lack these social cognitive abilities, then we would predict that people who have always found it difficult or impossible to believe in any gods might tend to be more male-brained. As male-brainedness appears to be in place at birth (though possibly amplified by subsequent decisions and interests), a resistance to theism could be in place at birth in such people. These are open research questions. Some evidence does exist, however, that atheists are much more likely to be male than female (in several studies, five or more times more likely).[8] And a well-replicated cross-culturally robust finding is that women tend to be more religiously involved and committed than men.[9]

It is tempting for those of us who are academics who notice that a much greater proportion of our esteemed colleagues are atheists than in the general public at large to conclude that being well educated or having superior intellect is an antidote against infection by religious belief. I take it that this is the point of my Oxford colleague Richard Dawkins's belief tabulation of eminent scientists in one of his books.[10] But as social scientists know, such apparent correlations must be handled with care as they are likely to have numerous potential confounding factors folded into them. For instance, psychologist David Myers has observed that Dawkins's atheist hall of fame is also essentially a white male club.[11] Could being white, male, highly educated, and an atheist have some other additional factor that relates these features and intelligence isn't the linchpin? It is certainly possible. These correlations between scientific achievement and atheism and between maleness and atheism provide us rather weak grounds for drawing causal connections between intelligence and atheism or scientific knowledge and atheism. What they do provide us with is additional motivation to explore whether something like Baron-Cohen's male-brainedness, which captures both a tendency toward elite ivory tower scientific success and maleness, could be an influencing factor in who turns out to be a thorough going atheist.

Religion scholar Benson Saler and colleague Charles Zigler have suggested another genetically linked personal factor that might indirectly contribute to atheism.[12] They offer that a particular set of genes may, through a specified set of neurochemical variations, make some individuals more or less anxiously on the lookout for the activity of intentional agency in their environment. That is, individual genetic inheritance may lead to variability in how likely individuals are to see events that others might consider a product of mere chance or natural processes as the activity of either human or superhuman activity. If evidence can be marshaled to support this claim in relation to belief in gods, this proposition would suggest that individuals who cannot understand why most people believe in gods might have inherited the genetic predisposition that makes them less likely to detect intentional agency around them. This genetic inheritance plus the male-brainedness, which appears to be (at least in part) a product of prebirth testosterone levels, could mark out individuals to be particularly attracted to atheism and mystified by theism.

Social-Environmental Factors

Personal factors may in part account for why there have always been a small number of atheists among human groups, but it may be prudent to look for environmental factors to account for the apparent rise in atheism over the past fifty years. Perhaps male-brainedness is becoming more common among the same populations that are (apparently) seeing a rise in atheism rates. Perhaps. But given that atheism appears to be clustered in a relatively small number of human environments, a combination of social and other environmental factors would likely be part of the answer.

Social Networks

William Bainbridge has suggested that one factor may be the web of intimate and dependent social relationships people have. He has suggested that people who actively care for others (traditionally, females more than males) cannot often fully satisfy the needs of others because some needs fall outside human power or resources. As a caregiver you may not be able to keep your children safe, well fed, healthy, and happy. Hence, suggests Bainbridge, caregivers resort to compensatory strategies, religious practice being among these. He writes, "Someone on whom no one else is dependent, someone who lacks strong social bonds of a kind to incur such obligations, is more free to espouse Atheism."[13] Consistent with this claim, Bainbridge shares survey data that show that single, childless, young males have much higher rates of atheism than others and having children works strongly against atheism. Provocatively, he shares data showing that societal atheism levels appear to be correlated with national fertility rates. Nations where people have fewer children (especially much of Europe) tend to have high atheism rates.[14]

Less Urgency, Fewer HADD Experiences

As I showed in Chapter 1, one factor that may contribute to belief in gods is the human cognitive system for detecting agents and agency around us. Dubbed the hypersensitive agency detection device (HADD), this system tells (under some conditions) that a self-propelled object is an agent, or creaking stairs are caused by an agent, or a patterned arrangement of rocks was created by an agent. Anthropologist Stewart Guthrie has argued that this tendency for detecting human-like intentional agency around us has been tuned by natural selection to register (at least fleetingly) the presence of agents given little provocation and even ambiguous evidence. Why? If HADD waited until the evidence for an agent nearby was conclusive, our ancestors might have ended up in the belly of a saber-toothed tiger. But if the wind causing rustling in the bushes made our ancestors bolt unnecessarily, not so much is lost. On balance, then. Guthrie argues, when in doubt, it is better to bet on an agent being present. He goes on to suggest that betting on unseen agents causing various natural events gets us well on our way to believing in gods.[15]

If HADD's hypersensitivity is tuned to survival, then it is likely to be more sensitive under conditions more closely tied to survival. That is, the more urgent it is to not miss the presence of an agent, the more likely it will be to register one. We are more likely to think we heard a person in the woods next to us when it is nighttime and we have heard that a serial killer is on the loose in the area than when it is midday and we are strolling though a family-friendly suburban park. When life is on the line, we would expect HADD to be more active than normal.

If the activity of HADD contributes to belief in gods and the urgency of the situation influences HADD's tendency to detect even unseen agency, then we would expect atheism to be more prevalent in environments in which urgency is low. Low urgency would mean a less touchy HADD. A less touchy HADD would mean fewer instances of detecting gods, ghosts, and the like.

Consistent with this analysis, atheism seems to be most widespread where urgency is relatively low: relatively affluent and safe postindustrial societies. Strict atheism is uncommon among hunter-gatherers, subsistence seafarers, and farmers.[16] Atheism is much more common where people do not have to worry about predators or prey, enemies stalking them, or storms that might destroy their crops. When food grows on supermarket shelves and protection comes from governmental agencies, HADD has much less work to do.

The Almighty Human

During 2007, the southeastern United States was afflicted by a severe drought, and the state of Georgia was particularly hard hit. The specter of completely running out of water loomed oppressively, and the state government wrangled with neighboring states and the U.S. government for water to be released from reservoirs upstream of Georgia. Emergency ordinances were passed by governmental bodies to restrict use of water. After months of pleas and negotiations with relevant entities, the governor of Georgia held a multifaith prayer meeting for people to ask their gods for rain. Not everyone regarded this as a productive activity, and the event was discussed and examined in popular media outlets. On *The View*, a popular morning talk show on television, the governor's prayer received a critical reaction from personality Joy Behar, who commented on November 14, 2007: "Well, they need to be praying to people who will fix global warming and take care of the environment. That's more realistic."

A generation or two ago, such a public sentiment would have been unheard of, not because of its apparent lack of faith in God's action, but because of its undue faith in human action. Behar's comments suggested that she regards humans as responsible for and capable of remedying natural disasters such as droughts. Such faith in human abilities, for good or ill, is a new phenomenon. Increasingly we expect that other humans, or at least human organizations and governments, can heal us of illness and cause and prevent flood, drought, famine, storms, fire, and pestilence. Science and technology have made these sentiments realistic in some cases. Humans can do amazing things. Such optimism in human power is especially prevalent in nations with strong governments, where the rhetoric of government being able to address all social ills, from poverty to happiness, is not uncommon (especially in election years). Add semisocialism to technological achievement, and it is not nearly so surprising that events and problems formerly regarded as acts of God are increasingly regarded as acts of humans.

Shortly after Hurricane Katrina devastated parts of the Caribbean and the Gulf Coast of America in 2005, I traveled to Belfast, Northern Ireland, to deliver a lecture. To my astonishment, all over walls and posts around Queen's University were notices for an upcoming event discussing how President George W. Bush *caused* Hurricane Katrina. Little did I know the American president had Godlike powers!

Such attributions of power to humans and human institutions has become credible because of scientific and technological advancements, and so when we are out of control and something striking happens that demands an explanation, rather than praise a god for deliverance or curse a god for calamity, another agency is capable of accepting the blame or praise: humanity. When HADD asks, "*Who* did that?" on the grand scale, "humans" increasingly appears reasonable as the answer—not because science has disproven all supernatural beings, but because science and technology have elevated collective human potential.

When the ten plagues befell Egypt, Pharaoh was eventually persuaded that a god was supporting Moses and that the Hebrews should be freed. Today Pharaoh might air-drop pesticides to kill off the locusts, flies, and gnats; demand a United Nations–led investigation into a neighboring nation's illegal dumping of slaughtered animal remains into the Nile, accounting for the amount of blood in the water; pass tighter emissions standards to try to clear the air of darkness; accuse Moses of biological warfare causing boils and human and livestock death; and lobby for a new international treaty for combating climate change because of the surprising amount of hail. A modern Pharaoh might not see the plagues as caused by a god at all, but as a combination of coincidence, poor environmental policy, and a well-coordinated and technologically savvy opposition-party conspiracy.

Pseudoagents Everywhere

Our strong tendency to see events as purposefully and intentionally caused by someone at times cannot be simply overridden as coincidence or blamed on a human being or group of people. Some events or situations look intentional and obviously fall outside the capability of people. Generally people have tended to see such events as the activity of the gods. The most obvious examples are the existence of the cosmos and the apparently purposeful design of things in the natural world from brains, eyes, and hands to entire animals and ecosystems. In recent decades, the apparently finely tuned properties of the universe as a whole, such that they make the emergence of carbon-based life forms possible, have generated strong impressions that the foundations of how everything works must have been done intentionally.[17] How does the ordinary atheist on the street satisfy himself that all of this apparent purposeful fine-tuning and design does not point to some kind of intentional agent? Enter the pseudoagents.

Fate, Destiny, and Providence all were mythologized into deities in the past. Today they still live on as pseudoagents that can explain handily striking events, fortune and misfortune. "Fate brought us together." "I guess it was destiny." "Providence is smiling on you." I call these *pseudo*agents because unlike proper agents, we generally do not think of these as having minds or mental states that direct their actions. Rather, they get plugged in as vague, stopgap explanations for events or situations that seem to demand an explanation because of apparent purposefulness, but a natural intentional agent (the first choice default explanation) and the conventional intuitive arsenal of physical and biological causation just will not do the trick. Suppose you lose your job unexpectedly one day, and the very next day an old friend, completely unaware that you just lost your job, contacts you with a new and better job opportunity. Speaking from experience, this is the type of purposeful-looking situation where your mind searches for an explanation for the coincidence. Can it be explained mechanistically? No. Biologically? No. By appeal to any natural agency? It sure does not look like it. So the drive to find a causal explanation arrives at some kind of superhuman agency—someone who could have known about the situation and coordinated the surprisingly favorable outcome. In a religious context, some kind of benevolent god would be a natural candidate. But if you are averse to the idea of gods, you need a different explanation. Providence would do nicely here.

Another pseudoagent, particularly popular among mathematically educated people including scientists, is Chance. If I flip a coin one hundred times and roughly fifty times it lands heads side up, I might say the number of time heads came up was due to chance. What "chance" is shorthand for in this situation is the claim that there really is not anything to explain, that what had happened would probabilistically be expected to happen based on ordinary natural laws continuing to function as they always do. So in such a situation saying, "It was chance," is not strictly an explanation but amounts to saying there is nothing to explain. Identifying what happens "just by chance" is such an important part of the sciences (including human and social sciences) and is invoked so frequently in reports, that Chance—with a capital C—can sometimes be mistakenly regarded as an explanation (as opposed to a claim that there is nothing to explain). So, someone might marvel that he won the lottery just a day after losing his job and want an explanation. If he wants to avoid appealing to God or Providence, he might say, "Aw, it was Chance." Maybe there really was nothing to explain in this case after all, but it seems that an explanation is needed. Chance sounds like the sort of explanation that educated people offer, so it will do. To take a more extreme illustration, if a person is brain dead for twenty-four hours and yet recovers, someone avoiding religious explanations might suggest Chance is a sufficient explanation for the event. Indeed, something like Chance can even do the job with the alleged fine-tuning of the universe. What if there were an enormous number of parallel universes? By Chance, there would be one that is suited for life, and, of course, we would have to be on it or we would not be here wondering about fine-tuning.

Where Fate, Destiny, Providence, Chance, and other pseudoagents fall short of supplying a satisfying explanation, a modern alternative is natural selection. From a naturalistic scientific perspective, natural selection is a purposeless, mindless process that happens to produce a diversity of living things such that they manage to survive and reproduce fairly well. But because natural selection serves to account for apparent design and purpose in the natural world (why an animal looks like this or why a biological process works like that) and we intuitively know that intentional agents bring about order, we frequently mistake natural selection for an intentional agent, speaking of what it "designs" or "gives" or "cares about" or remarking that it is "blind" or "cruel."[18] As I sketched in Chapter 3, evidence exists that children find evolution by natural selection relatively difficult to adopt over and against creationist explanations for animals. Nevertheless, natural selection can play the psychological function of an intentional agent concept when trying to explain elements of living systems.

Natural selection is an abstraction (a theoretical process) that sometimes gets used as a purposeful intentional agent to explain states of affairs in the natural world that seem to demand a mind behind them. Other abstractions manage to take the place of appeals to divine agency or order as well. For instance, it is not uncommon to hear "It's the *government's* fault" that something is unsatisfactory or something bad has happened, as in "It's the government's fault that my children are receiving a poor education." In such a statement, government is the agent responsible for a wrong. Similarly, the economy or the market might be blamed for my losing a job. Instead of blaming specific humans or divinities for a problem (or praising them for a benefit), an abstraction (the market) is inserted as the cause, even though it might be difficult to specify how exactly the market (whatever that is) causes me to lose my job. On the positive side, we might hear how liberty demands a certain course of action or justice triumphed to right a wrong. Liberty and justice are abstractions that get spoken of as agents. Not surprisingly, people frequently personify governments, markets, liberty, justice, and other abstractions as in Uncle Sam for the U.S. government, the invisible hand of free-market capitalism, Lady Liberty, and the blindfolded Lady of Justice.

I am not arguing whether these alternative explanations that might be used in a god's place are good or bad explanations, or if they really amount to mutually exclusive alternatives for the activity of a god or gods (I suspect they do not). Rather, I have listed some examples of the sorts of reasoning that, if collectively encouraged, can help stymie natural dispositions toward supernaturalism. Not logically but psychologically, having a stable of different explanations (including natural agents and pseudoagents) to plug in when intuitive causal cognition demands an explanation and is inclined toward accepting supernatural agency keeps supernaturalism less relevant and, hence, less credible.

Time for Reflection

A final class of social-environmental factors that might help people resist the natural pull of supernatural thinking is time for reflection. With any natural cognitive dispositions such as those that govern our intuitive thinking about physical objects, living things, mental states, languages, morality, and so forth, we need more time, attention, and effort to override these dispositions than to yield to them. For scientific reasoning to break free of the constraints of natural causal reasoning, slowed-down, rarified situations to think carefully are required, often with cognitive props such as writing systems and mathematical codes. Similarly, for theologians to develop more sophisticated belief systems that deviate from natural religion grounded in ordinary cognitive biases, time for dispassionate reflection is required (see Chapter 6). Such intellectual work is not easy (but may get easier with practice), and so time and concentration are helpful. Atheism in a strict sense may not be as cognitively unnatural as modern science, but it is more unnatural than general religious thinking. As such, time for reflective reasoning and a culture that values and supports such pursuits would be helpful in atheism's maintenance.

Early Instruction

I bumped into a colleague in the British Museum, almost literally. She was distractedly pushing around a stroller with her three-year-old son in it, and we nearly collided. As good mothers do, she was pointing out exhibits and enthusiastically explaining them to her adorable tyke. I was fortunate enough to overhear her reinforcing what was evidently a common refrain in her home about how natural selection brought about the animal diversity we see today. With the right educational techniques, perhaps particularly applied early in development when children's cognitive equipment is still very flexible, children might be taught evolution by natural selection even though it is cognitively challenging and somewhat unnatural. Perhaps, too, if instructed early and often while their cognitive systems are still shaping, children might be discouraged from connecting perceived design in the natural world to gods.

I have argued that a steady regime of religious indoctrination is insufficient to account for the global and panhistorical commonness of religious thought. People seem to have natural tendencies to religious thought and action, but this does not mean that a steady dose of counterreligious indoctrination might not be helpful in fighting against these natural tendencies, much in the same way that regular practice with statistical or scientific reasoning improves our abilities to think statistically or scientifically, even though such thought is relatively unnatural.

This counterreligious indoctrination could include offering children alternative ways to interpret their perception of design and purpose in the natural world and their detections of nonhuman agency. Chance, government, natural selection (as an intentional, directional agent of change), and other pseudo-agents could do nicely here.

How to Be a Confident Atheist

Religious belief is relatively widespread in people in part because early-developing conceptual tendencies predispose children to understand and accept religious ideas. Certainly testimony, social, and other environmental influences contribute to whether children will become true believers in a particular religious tradition or become agnostics or atheists. But when atheism does battle with supernaturalism over the hearts and minds of people, the playing field is not level from the beginning.

Additional personal or social-environmental factors need to be in place to help atheism survive and spread. So if you want to be a satisfied, confident atheist, my best advice is to follow these seven steps:

1. *Have less-than-average fluency in reasoning about minds or a tendency to not use psychological or social reasoning.* Unease with which one employs intentional agent explanations may dampen religious thought.
2. *Do not have children.* Living with few dependent relationships may enable unbelief.
3. *Stay safe.* Environments in which one's survival is rarely under threat and survival-relevant decisions are rarely made may reduce gravitation toward religious entities. Avoid having to hunt to survive or to scratch out an existence from unpredictable earth and weather.
4. *Get in the habit of crediting or blaming humans for whatever you can.* If you detect intentional agency in the environment, when human activity can be a useful explanation for this agency, as is often the case in urbanized societies especially with strong governments and technological sophistication, the need to turn to gods to account for perceived intentional agency is limited.
5. *Learn to like pseudoagents.* For those situations and occasions that clearly fall outside ordinary human

control, if pseudoagents and abstractions such as fate, chance, justice, government, and natural selection are available as explanations reinforced by common discourse, then gods can be better pushed aside.

6. *Take time to reflect.* A key ingredient for resisting the naturalness of religious thought is time to dispassionately reflect. Just as theologies that deviate from natural conceptual biases need reflection time to be worked out, so too nonreligious philosophies and worldviews need reflection.
7. *Add to these factors indoctrination of the young against religion.* Then nonbelief might stand a fighting chance.

Notice that these factors that I am suggesting contribute to the spread of atheism are relatively rare in human history but are increasingly common, particularly in European societies.

TEN

Should You Introduce Children to God?

How to teach children about God, what to teach, and when to teach it are all challenges that many parents, religious leaders, and religious educators face. The scientific research concerning children's cognitive abilities relevant for religious thought certainly informs these questions. In the concluding chapter, I build on what the evidence seems to be telling us about children as born believers and offer some tentative suggestions about what it all means to teaching kids about God. But before turning to these matters, in this chapter I briefly address whether people *should* teach their children about God at all. After all, some have asserted that teaching children to believe in God is wrong—even child abuse. In a chapter titled "Is Religion Child Abuse?" Christopher Hitchens's sentiments about the clergy are captured in this way: "For centuries, grown men have been paid to frighten children in this way (and to torture and beat and violate them as well, as they also did in [James] Joyce's memory and the memory of countless others)."[1] Similarly, in reference to another anecdote about informing a child about hell, Hitchens writes, "Those who lie to the young in this way are wicked in the extreme."[2] Hitchens is not alone or the first to suggest that teaching children to believe in a god is abusive and should be stopped.

Respected scholars have even contemplated criminalizing the teaching of religious beliefs to children. Consider the following words from British psychologist Nicholas Humphrey's Amnesty Lecture:

> Children, I'll argue, have a human right not to have their minds crippled by exposure to other people's bad ideas—no matter *who* these other people are. Parents, correspondingly, have no god-given license to enculturate their children in whatever ways they personally choose: no right to limit the horizons of their children's knowledge, to bring them up in an atmosphere of dogma and superstition, or to insist they follow the straight and narrow paths of their own faith.
>
> In short, children have a right not to have their minds addled by nonsense. And we as a society have a duty to protect them from it. So we should no more allow parents to teach their children to believe, for example, in the literal truth of the Bible, or that the planets rule their lives,

> than we should allow parents to knock their children's teeth out or lock them in a dungeon.[3]

Humphreys, Dawkins, Hitchens, and others of their ilk genuinely and sincerely believe that teaching your own children to be religious believers is a form of child abuse. These men (all whom I know of who take such a line are men) are not stupid or mad, but are honestly working out the implications of their worldview.

Though stopping short of claiming society's duty to protect children from their parents' belief systems as Humphrey argues, Dawkins does see enculturating others with religious beliefs as comparable to physical or sexual abuse. In a recent book, Dawkins writes:

> Once, in the question time after a lecture in Dublin, I was asked what I thought about the widely publicized cases of sexual abuse by Catholic priests in Ireland. I replied that, horrible as sexual abuse no doubt was, the damage was arguably less than the long-term psychological damage inflicted by bringing the child up Catholic in the first place.[4]

Dawkins goes as far as to claim that referring to a child of Christian parents as a "Christian child" or a child of Muslim parents as a "Muslim child" is abusive: "Even without physical abduction, isn't it always a form of child abuse to label children as possessors of beliefs that they are too young to have thought about?"[5] An inaccurate labeling, perhaps, but abusive? Even if the child self-labels in this way?

Again, part of me wants to give Dawkins the benefit of the doubt and believe he is using overstatement to make a point, but his comparison between religious upbringing and physical and sexual child abuse is too lengthy and developed to be easily dismissed as rhetorical bluster. Dawkins claims that raising sensitivity about inappropriately labeling children as members of a particular religious group is his primary aim in his chapter titled "Childhood, Abuse and the Escape from Religion." I would be much more convinced that this was indeed his primary aim had he spent time discussing such related matters as at what age children are old enough to label themselves and whether such labels are appropriate when they are also cultural or ethnic markers. For instance, the terms *Hindu* and *Jewish* carry ethnic as well as religious meaning—a dual role found throughout the world where to be the member of a cultural group is to be a member of the religious group and vice versa. I wonder, too, how a parent convinced of Dawkins's point about hasty labeling answers her child, "Mommy, if you are a Mormon, why can't I be one too?" Rather than take up these issues to fortify his position, he spends considerable time on the more provocative matter of comparing biblical teaching with sexual abuse.

After sharing an anecdote about a girl who reportedly found being taught about hell as more psychologically tormenting than being fondled by a priest, Dawkins writes: "I am persuaded that the phrase 'child abuse' is no exaggeration when used to describe what teachers and priests are doing to children whom they encourage to believe in something like the punishment of unshriven mortal sins in an eternal hell."[6] These comments offer a scientifically testable hypothesis: Does being taught the existence of an eternal hell during childhood cause comparable (or more) psychological harm than being physically or sexually abused? To my knowledge, this particular comparison between the psychological consequences of childhood sexual abuse and childhood teaching about hell has not been conducted. I am not, however, optimistic that the relevant evidence would support Dawkins's position.

Considerable research on the relationship between religious commitment and psychological and physical well-being does exist, and the general finding is that committed theists are psychologically healthier and more equipped to cope with emotional and health problems than nonbelievers. As the vast majority of this research has concerned predominantly Christian populations for whom learning about hell was part of their upbringing, it at least suggests that hell and other "abusive" doctrines do not succeed at wiping out any benefits of religious participation. The negative consequences of childhood sexual abuse likewise are well documented.[7] Taken together, without more direct systematic evidence, it is irresponsible for an informed scientist to approximate religious upbringing with sexual abuse.

Dawkins and Hitchens both unscientifically present the same kind of anecdotal evidence in support of their claim, a case of someone who remembers being traumatized by learning about an eternal fiery hell. If we are swapping convenient anecdotes, I could share that I too have unpleasant childhood memories of learning about sin, punishment, and hell. When I heard the passage of the New Testament in which the merciful and compassionate Jesus reportedly said, "But I say to you that everyone who is angry with his brother shall be guilty before the court; and whoever shall say to his brother, 'Raca,' [empty-head] shall be guilty before the supreme court; and whoever shall say, 'You fool,' shall be guilty enough to go into the fiery hell,"[8] I knew I was in trouble. I called my older brother a "fool" all the time. I felt guilt, anxiety, and dread. I also remember once being frustrated by getting a toy stuck in a tree and letting fly a particularly strong exclamation about the tree and then being devastated by the realization that I had "taken the Lord's name in vain." I remembered reading in the Bible that "you shall not take the name of the Lord your God in vain, for the Lord will not leave him unpunished who takes His name in vain."[9] What kind of punishment? Perhaps a dip in the lake of fire? Certainly scary stuff. But we have no evidence that in even a sizable minority of children are these episodes so troubling as to justify Hitchens's comment that "those who lie to the young in this way are wicked in the extreme." In my case, these events led to my being motivated to be kinder to my brother, to better control my tongue, and to understand that even though I was a basically good kid, I was just a sinner like everyone else—no better—and needing God's mercy. I would trade a little childhood angst for these outcomes all over again. I was also deeply troubled by learning about the Holocaust and the way images of emaciated, fly-covered, rotting piles of bodies seared into my young brain and disturbed my sleeping and waking thoughts for some time. Indeed, images of the Holocaust haunt me to this day more than images of hell. Was it wicked or abusive for me to be taught about the Holocaust? Hitchens might answer that this is different because the Holocaust was real and not a lie, but such an answer would surely play into the hands of the fundamentalists he abhors. "Well, hell is real, too, and not a lie," I can hear them saying. This argument equates teaching falsehoods with child abuse, while teaching truths, however disturbing, gets a free pass. The disagreement, then, reverts back to a concern over what is or is not true, a disagreement not resolved by accusations that the teaching is abusive or "wicked in the extreme."

More modestly Dawkins proposes what moral parents and guardians should be offering their children instead of indoctrination: "I thank my own parents for taking the view that children should be taught not so much *what* to think as *how* to think."[10] I am sympathetic to this sentiment: teach children how to think well and how to learn for themselves, and you equip them to cope independently with an ever-changing world.

Difficulties arise, however, when trying to teach young only how to think without teaching them what to think—particularly very young children in the real-world pragmatics of curious child-parent exchanges. How do parents answer their children's queries such as, "What happens to people after they die?" and "Where was I before I was born?" and "Why do you pray?" without at least hinting at her own beliefs on such matters? In a classroom of college students, it might be possible to explain that some people think this for such-and-such a reason, but other people think that for such-and-such a reason, and conclude with, "But you have to decide for yourself." Even children who do not immediately ask, "But what do *you* think, Mommy?" are bound to deduce from Mommy's tone, her reactions to the child's own thoughts, and her day-to-day behaviors and overheard conversations with other adults just what Mommy really believes. The ideal of not directly or indirectly teaching your moral or metaphysical values and commitments to your children is sheer fantasy, and the attempt could prove emotionally and psychologically harmful. Would a child feel loved and affirmed if her parents refuse to tell her what they believe in answer to her most heartfelt questions?

I had an exchange with one of my own children (then at age eleven) that highlights these difficulties. My older child had asked about the beliefs of various groups: Baptists, Catholics, Lutherans, and so forth. I tried to give a fairly factual and even-handed appraisal of these worldviews. My eleven-year-old interjected, "What am I?" I explained that she had to decide that question for herself. I imagine Professor Dawkins would have been pleased with me, but then my daughter insisted, "But what am I?" I again tried to explain that she had to consider things for herself and make that decision when she was old enough. Somewhat irritated with me, she persisted, "I'm a girl but what are *we*?" As we talked, it became clear, and I felt obliged to answer, what it was that I considered *myself* because that is how she wanted to identify *herself.* So I asked, "You want me to tell you what we [indicating her mother and me] are so you can be what we are?" She replied, "Well, yeah." Such an exchange captures a typical dynamic. When it comes to determining what children think, believe, and value and how they identify themselves, social considerations can be as important as intellectual ones. Children (generally) want to be what their parents are, on any number of religious and nonreligious dimensions, and to disallow them inclusion in that social circle is a form of emotional exile. Identifying with their parents is natural.

I fear that Humphrey's and Dawkins's strategy of parents' not providing children with a thought-through belief system (religious or otherwise) is pragmatically doomed and more likely than not to create relational distance and estrangement within families. I suspect a child systematically excluded from his or her parents' values, beliefs, practices, and social identification, if it were truly possible, would be more likely to feel "abused" than a child who is taught to believe what his or her parents sincerely believe. Such a child will also more likely gravitate toward unrefined natural religion.

I agree with Dawkins that parents should, as much as children are capable, help them to learn how to think as opposed to simply telling them what to think. Unfortunately, Dawkins amplifies his position in a problematic way. Approvingly he writes, "Humphrey suggests that, as long as children are young, vulnerable and in need of protection, truly moral guardianship shows itself in an honest attempt to second-guess what they *would* choose for themselves if they were old enough to do so."[11] Humphrey argues that "the only circumstances under which it should be morally acceptable to impose a particular way of thinking on children, is when the result will be that later in life they come to hold beliefs that they would have chosen anyway, no matter what alternative beliefs they were exposed to."[12] This perplexing strategy of trying to imagine what a child would choose for herself if old enough runs aground in at least two ways.

First, this strategy asks us parents and guardians to imagine our own children at an age at which they would competently choose for themselves what they should be doing or believing now, a difficult feat for the imagination. To illustrate, if I am a parent of a twelve-year-old, should I try to "impose" sexual abstinence on my child (or at least attitudes against casual sex at age twelve)? To decide this, I must imagine a time when my child is old enough to make the decision as to whether his or her twelve-year-old self would be old enough to make such a decision. At what age should I imagine my child being? A fifteen-year-old who perhaps does not understand any health, safety, or social risks? Or perhaps a nineteen-year-old who understands the risks but sees social consequences here and now as more important than potential health risks later? Or perhaps a twenty-six-year-old who understands the value of delayed gratification? It is not clear how old a child needs to be to make informed decisions of various sorts.

This problem dovetails with a second one concerning how we go about deciding what it is our hypothetical child would decide for herself or himself. Being in a position to make good decisions has more to do with experiences than simple age and includes the cumulative experiences that resulted from parental decisions—decisions like the one we are trying to make by imagining what decision the child would make. Which potential, hypothetical version of our child should we imagine making a decision for his or her younger self? In considering my twelve-year-old's sexual behavior, I could imagine a twenty-six-year-old version who chose to abstain and was glad (or sorry) that she did so; or a twenty-six-year-old version who chose to freely experiment, contracted sexually transmitted diseases, became pregnant at fourteen, never completed school, lives in poverty, and has great regrets; or a twenty-six-year-old version who chose to freely experiment and managed to avoid pregnancy and disease, and now finds her previous sexual exploits gratifying. Which shall I imagine? In a similar way, Dawkins encourages us to imagine whether an Amish child would choose to grow up Amish if old enough to decide for himself or herself. Well, which version of the possible older child shall we imagine: the one who was not raised Amish or the one who was? Presumably Dawkins (and Humphrey) want us to assume the former, non-Amish upbringing. After all, who would choose to be raised Amish? Right? Such thinking stinks of egocentric and ethnocentric xenophobia. Deciding what children *would* choose for themselves (were they old enough to choose) is an exercise in fanciful speculation and hardly a strong foundation for a moral parenting strategy.

I suspect the error the new atheists detect imperfectly is not parents' lovingly sharing with their children the deepest commitments and values of their hearts, but making love for their children conditional on accepting and agreeing with those commitments. Certainly parents who withdraw love or even threaten physical punishment for children who disagree with their worldview can constitute a form of abuse and represents a moral failing on their part. Nevertheless, I see no reason to suppose that religious parents wishing for their children to be religious are any more prone to perpetrate such injustice than atheist parents wishing for their children to be atheists. Whether parents want the responsibility, and whether outsiders want some parents to have the responsibility, it is up to parents to seek what is best for their children and make decisions with their children's best in mind. These decisions will inevitably be colored by the parents' cultural perspectives, including their religious or nonreligious worldviews.

The new atheists too are concerned (rightly, I think) with

the continued propagation of false ideas in the name of free religious education. Parents teach their children nonsense all the time, and often for very poor reasons. Worse still is when parents teach their children ill-conceived notions and then go on to discourage children from critically examining those notions. Some of these faulty ideas are religious; many are not. We should value the pursuit of truth, wherever reason and evidence lead, and hope that parents and educators will not erect unnecessary barriers for children to be able to discern truth, whether in a religious sphere or not. Why should an idea get a pass regarding proper ways of discovering truth just because it is "religious"?[13] As philosopher Roger Trigg comments, "The right to religious liberty, however precious, can never be so unqualified as to leave the lives of young people at the mercy of those who have contempt for the voice of reason."[14] If parents believe their religious commitments to be reasonable and true, encouraging children to discover the truth should not be threatening.

If parents have considered their own religious beliefs and regard them as true and important, I see no justification for condemning parents who encourage their children to adopt those religious beliefs as "immoral" or "abusive," unless they do so in an unloving way. Again, Trigg writes, "When religion cannot be transmitted to the next generation, it is not being freely exercised."[15] I am tempted to defend atheist parents' passing on their atheist beliefs in the same way, but must mention one wrinkle: scientific evidence demonstrates the personal value in committed religious practice.

Though there are many spectacular examples to the contrary, research does indicate that commitment to a religious belief system and participation in a religious community is associated with many positive outcomes. Actively religious people have been shown to enjoy more mental and emotional health, recover from trauma more quickly, have longer and happier lives, are more generous, volunteer more, and actively contribute to communities more than nominally religious or nonreligious people do.[16]

Why religious participation has these positive outcomes is still an active area of research, but one way religious beliefs—independent of being part of a nurturing community—can operate in a way that produces personal well-being is by serving as a meaning-making and organizing principle for our various goals and motivations.

We all have numerous day-to-day goals, or *strivings*. We strive to be popular with our peers or coworkers, to look intelligent, to work hard, to make our spouse happy, to be a good parent, to live a healthy lifestyle, or whatever else our personal strivings may be. To discover some of your own strivings, simply complete this sentence: "I typically try to . . ." Strivings vary from person to person. Research on strivings by psychologist Robert Emmons and his collaborators has shown that people who have a high proportion of strivings having to do with God or with their religion (such as striving to be a good Muslim or to please God) tend to have better physical and mental health.[17] University students with relatively high numbers of religious strivings reported lower incidences of depression, anxiety, and other common emotional distress, as well as fewer bouts of illness and visits to the university health center. It seems, then, that religious strivings are related to health, but why?

Emmons and his team asked college students to generate a

list of personal strivings and then had them score those strivings against each other in terms of how much they helped or hindered the satisfaction of each other. For instance, if one of my strivings is to be careful with money and another is to live every day as if it is my last, these two strivings would be in conflict, hindering each other, whereas "living a simple lifestyle" as a striving might be in harmony with the striving to be careful with money. Scoring the amount of conflict between each pairing of strivings yields an overall conflict score. It turns out, perhaps unsurprisingly, that people with a high degree of conflict among their strivings find life stressful and are more prone to anxiety, depression, and physical illness. So, like high numbers of religious strivings, low conflict among strivings is associated with better well-being. Could low conflict among strivings and religious themes in strivings be related to each other?

Emmons's research has shown that people with lots of religious strivings do tend to have less conflict among their strivings. Rather than being torn in every which way by competing values and desires, religiously centered people tend to find that their strivings work together. Three separate strivings such as "being a good citizen," "being a good friend," and "pleasing God" could all be seen as in harmony and meaningfully related to each other. Being actively religious, to the point that religious beliefs have an impact on one's day-to-day goals or strivings, promotes well-being by structuring and ordering what is important in life, thereby reducing conflict, and thereby reducing mental and physical illness.

Undoubtedly additional factors contribute to the fact that religiousness, including the content of particular religious beliefs and community participation, can exert a positive and constructive influence in individuals' lives. I share Emmons's research because it suggests one pathway that might be especially important in late childhood and the adolescent years in harnessing greater reflective reasoning skills to begin connecting basic religious beliefs to day-to-day life. If this can happen, if "There is a God" can be connected with "I should exhibit integrity in all that I do" or "I should put others' needs before my own" or the like, then a life of thriving may result.

That some religious beliefs when put into action seem to promote positive mental, emotional, and physical health as well as the exercise of character strengths does not mean that these religious beliefs are necessarily true (or false). We cannot equate *beneficial* or *useful* with *true.* If a fire was to break out in your kitchen without your knowing but (for some odd reason) you thought a hungry bear was in there, you would probably run to safety. The belief that a bear was in your kitchen would be useful and beneficial to you in that context but still be false.

If you believe in a good God who made the world and the people in it, then you might find it surprising if being committed to God had absolutely no benefits in the here-and-now. Nevertheless, just the presence of benefits does not count as justification for believing in God.

Religious commitments generally appear to promote psychological and physical well-being, provide tools for coping with life's challenges, and can offer a meaningful framework for integrating goals. Given these facts, is it moral for a parent to actively discourage a child from beliefs that carry these benefits? Perhaps so, provided that in the parents' worldview, the pursuit of truth justifiably trumps individual human flourishing, but this is a decision for parents to thoughtfully make. The parents who regard their religious commitments as true and beneficial are justified in lovingly and thoroughly instructing their children in the ways of their religion. In the next chapter, I offer some suggestions for how to do so effectively.

ELEVEN

Encouraging Children's Religious Development

A CHRISTIAN MOTHER shared this illustrative story with me:

When Ben was in kindergarten we were home schooling and we were doing our Bible lesson, the gist of which was to study the life of Jesus and try to live like He did. So at the end of the lesson I made a generalization that being a Christian is simply being like Jesus. Ben says, "I don't have to." I said, "If you are a Christian that is what you should try to do." He said, "No, I don't." After some back and forth and Mother's fear that her child was choosing to

> spend eternity separated from God at the age of 6, I asked "What is it you are supposed to do?" Ben's reply? "I just watch Daddy—he has to live like Jesus and I have to live like him."

This story captures several truths. First, even six-year-olds can be thoughtful religious believers. Second, young children do not slavishly agree with everything their parents tell them. Third, when it comes time to encourage or impart information, the behavioral model of a parent or teacher means a whole lot more than doctrine or Bible stories.

Throughout this book I have argued that the available evidence points toward children being naturally receptive to basic religious beliefs such as believing in the existence of gods. You might think that this naturalness means there is no reason to try to teach children. Won't they just find religion for themselves? Unless special environmental or personal conditions of the sort that I sketched in the previous chapter get in the way, children will become religious without any direct instruction or teaching. But the naturalness of religion does not imply that any religious beliefs are just as good or just as true as any other. Given the relative importance of religious beliefs for motivating life decisions and contributing to health and well-being, I suspect most adults would want the children in their care to have the best they can offer. As I noted in Chapter 6, the sort of religious beliefs children naturally acquire without any explicit input from adults will deviate from the worked-out systems of theology of the world's religious traditions. Left to their own devices, they will likely become religious in some sense but probably in a sense more like what you would call superstition than a thoughtful, sophisticated belief and behavior system. They may be drawn to worshipping Mother Earth, astrology, or an unhealthy preoccupation with ghosts, among other suspect beliefs and practices such as wearing their underwear inside out to produce snow or carrying amulets for success on school exams.

Failure to provide children with a system of beliefs and practices refined and honed through generations of theological debate, philosophical scrutiny, and collective experience may lay them open to religious extremism, catchy anecdotes from charismatic spiritualists, or buffet-style picking and choosing the most palatable options from various available traditions, whether or not they actually make sense together. Children's natural propensities toward religious thought and hunger for spiritual fulfillment will propel them toward some kind of religious expression whether trusted adults supply suitable targets or not.

Compare religious thought to food preferences. Left on their own, children will gravitate toward foods that will keep them alive, but they may not be the best foods for a healthy life. Human biology and psychology restrict the sorts of things we are tempted to eat and will find palatable and digestible. It may take some upset stomach and surviving some occasional bouts of poisoning, but eventually children will discover what they can and cannot eat, and they will discover junk food. Their diet may consist of as many fats and sugars as possible. They may eat more like Homer Simpson than Emeril Lagasse or Jamie Oliver. Nature will help children find foods to survive, but it takes the guidance of parents and others to help kids discern what balance of foods is best for them nutritionally and how to prepare and eat food so that it is not just a source of calories but a more nutritious, sophisticated, and pleasurable experience. Similarly, on their own, children will tend to become religious, but not necessarily in the best, most reasonable, or most beneficial type of religiousness.

In Chapter 9 I offered some suggestions for those hoping to keep atheism in the face of natural tendencies toward supernatural belief. Here I offer some suggestions for those who wish to encourage children's religious beliefs. I begin with a caution: because participation in a religion can powerfully shape a person's beliefs and values and can motivate people to both striking acts of benevolence or malevolence, I suggest parents examine the justifications for their own religious beliefs and commitments before blandly encouraging their children to own these beliefs. Parents should consider their reasons for beliefs of any kind, but perhaps especially those that may play a powerful role in their children's lives if passed on. What I advocate is not a cynical stance but a humble one. We could be wrong about our commitments and should welcome what intellectual (including religious) traditions, the exercise of reason, and scientific evidence have to offer us by way of challenge, correction, or affirmation of our beliefs. Then when parents turn to offer insights to their children, they can feel confident that they have performed quality control on their own beliefs and are offering their children the best they have to offer. As a parent, I try to give my children skills for discerning the good and true in hopes that they will adopt my right beliefs but also reject my mistakes. With this caution, I offer some summary thoughts on how to encourage children's religious development.

1. *Start early.* Children can handle thinking about gods and other religious ideas from very early ages. You need not turn an almighty God into a cartooned bearded man for the sake of three-year-olds, but you can teach about divine attributes such as being superknowing, superperceiving, immortal, and wholly good. In fact, it may be that starting children when they are three may be more effective than waiting until they are eight or older. Just avoid abstract and complex language, and give children tangible thought problems through which they can exercise their understanding. Many of these could be patterned after the experiments I have shared in earlier chapters. For instance, instead of merely telling kids about God's ability to see or hear everything, give them games in which they speculate about what God can see or hear versus what a person or various animals can see and hear. "Can God see what is in this darkened box?" "That's right! God can. But could your brother? Could a dog?" Instead of only telling them that God is immortal, also ask children questions about what God was like before they were born, or what God will be like many many years from now. Was there ever a time when God did not exist? Did God need to be born? Will God ever die?

Young children are especially ready to grasp the idea of a divine creator of the natural world. You could draw children's attention to what they already are inclined to see: the apparent function and purpose in the natural world. Then answer this question: *But where does the design come from*? Don't forget that children can see the sun, moon, mountains, lakes, rivers, and rock formations as functional and purposeful. When it comes to living things, you, like a number of contemporary theists, might want children to understand the cosmos as created by the divine, but you also want them to understand evolution by natural selection as the process through which God brought about the diversity of life. Though research suggests evolution may be difficult for children to accept, it might actually be easier if children see it as having divine muscle behind it to account for the seeming directedness of evolution.

2. *Teach children what you regard to be true with love and humility.* Whether you want to or not, you will teach your children what you regard as true about the world, life, relationships, values, morality, and so on. Rather than shy away from this responsibility, use your job as teacher and role model as impetus to carefully examine your own beliefs first and then share with the children in your care what you regard as good and true, but always with a loving spirit of confident humility. Examine your motivations behind the teaching. Are you acting in love, trying to help your children develop into fulfilled and effective adults, or are you acting selfishly, to make your children think and behave just like you?[1] If you are motivated by love and have examined your commitments, be confident in your teaching because you have weighed your beliefs carefully. Nevertheless, practice humility because you could still be mistaken, if only in small ways.

Part of religious education is self-discovery, but only part. As philosopher Roger Trigg highlights repeatedly in his book *Religion in Public Life*, often forgotten in discussions about children's religious education is that religion is not merely an exercise in self-discovery or finding meaning for oneself.[2] Religious education also concerns claims about how the world, humans,

adult testimony, he and his collaborators offer a number of strategies that they suspect contribute to children confidently believing in a particular entity. Paraphrasing and applying them to teaching about God in particular, here are three of his suggestions.

4. *Don't say you believe in it or have faith in it; talk as if there is no question about it.* Harris notes that when people talk about germs or oxygen, they do not typically say, "I believe in oxygen. I believe that oxygen is here around us." Similarly, people do not claim to believe in chairs, cats, shoes, or the sun. In contrast, people do talk about believing in God, spirits, angels, and so forth. Children might be sensitive to the way "believing in" is usually used for entities in which doubt of existence is possible. Similarly, we do not typically say we "have faith in" the chair we sit on or the car we drive, even if we do. Consequently, this sort of talk may set apart God, spirits, and angels as more tentative than germs or oxygen. It may be appropriate on occasion to address the fact that others have divergent views. Scientists and other scholars do this with regard to cutting-edge areas of research where consensus is still being forged. But rather than say, "I *believe* Y," the confident scholar says, "Others *believe* X, but Y *is* the case, and here is why."

When talking about "believing in God" or having "faith in God," adults do not simply mean they believe that God exists. They may also mean that they trust in God the same way we "have faith in" our friends or spouses or when we tell our children, "I believe in you." If this relational trust is what you mean by saying you "believe in God" or "have faith in God," do not be vague. Use less ambiguous language and say, "I trust God."

5. *Talk about God in actual contexts in which God's action can be detected.* As an illustration for a family sermon, my church vicar began by opening presents and pretending that they were one thing when they were another. For instance, he unwrapped a frying pan and said, "Oh! It's a ukulele. I've always wanted to play a ukulele!" And then he began strumming the pan and singing. To the amusement of the adults, a child sitting in the front shouted repeatedly, "No! It's a frying pan!" Children—even three- and four-year-olds—are very sensitive to the difference between reality and fantasy, but some people are tempted to talk about God and religious matters only in rarified, special contexts that may encourage thinking of God or spirits as part of the fantasy realm. Unlike talk about fairy godmothers and leprechauns, talk about oxygen and germs more frequently takes place in explanatory contexts, that is, explaining why something is the way it is. You feel funny underwater because you cannot get any oxygen. You get sick because of germs. Likewise, to group God with these other unseen forces, you could say that a person gets well because God healed her. Talk about fairy godmothers may be causal but not in real-world contexts. They turned pumpkins into coaches "once upon a time," not yesterday in my backyard.

Harris suggests that incorporating forces and agents into cause-and-effect relationships—the sorts of reasoning that children naturally pay considerable attention to—may be more effective in fostering commitment than abstractly postulating a thing's existence. The lesson applies to those teaching science as well as religion. Use the ideas in cause-and-effect contexts if you want them to seem worth believing in and believable. Imagine an elementary school science class in which the teacher merely posited the existence of invisible subatomic particles called "quarks" that come in six flavors. The nerdiest of the class may find this claim interesting, but most will go glazed-eyed because they will not know what follows from the existence of quarks. What do quarks do? How would things be different without them or if they were different? Children's minds are drawn to trying to figure out how things work, how they feature in causal relationships, especially those that might matter to the child. Bicycles, dolls, shoes, and pizzas matter more to children than neutrinos, black holes, and trilobites.

With this principle in mind, talking about God's actions in cause-and-effect contexts in the here and now will be more effective than talking about God in the abstract or even what God did at the creation of the world, with Noah and the Flood or with Moses and the Exodus. These sorts of stories alone may give children the impression that God is more like a character in fantasy tales about fairies and ghouls than a real player in the world. If you believe God heals relationships and bodies, helps change people's minds, or finds you a parking space, say so in earshot of your children.

The hypersensitive agency detection device (HADD; see Chapter 1) needs to be able to periodically link events in the world with the action of gods, or the gods will become less relevant. If God is talked about only in terms of things God once did during a mythical age, God's actions are unlikely to be noticed here and now. Prayer, particularly asking and thanking God for common things like family members, health, and wealth, as well as mundane events such as getting a good deal on tomatoes at the market or having a particularly enjoyable outing with friends, could help prime the one who prays to notice God's actions. Of course, noticing God acting or even wondering at why God did not act as requested (Was there a greater good at stake? Was the request appropriate?) encourages bringing thought about God into causal connection with the real world.

You might be worried that God has competition in these explanatory roles: medicine, not God, heals us; the bad weather was not a punishment but the result of a passing storm system; and the parking place opened up because of dumb luck. Science and more sophisticated probabilistic reasoning have pushed God's activities to the margins of day-to-day activities. It is not my place here to develop a theology of divine action, but let me note that one kind of explanation for an event need not compete with another sort of explanation. I may have fallen to the ground because of gravity, or a failure of my cerebellum and motor cortex to process signals from my vestibular system in time for me to recover my balance, or because someone left a toy car at the foot of the stairs. All three explanations can be relevant at the same time. Similarly, I can be simultaneously sick because of germs and because an evil spirit used germs to infect me. Naturalistic or scientific explanations for events need not compete with or eliminate religious explanations. Multiple explanations may be true and helpful simultaneously. Hence, I am not suggesting that God or other religious forces replace scientific explanations but rather that they add ultimate causes behind the immediate causes.[5]

6. *Use religious ideas in mundane circumstances and not just special ones.* An extension of the previous suggestion is to connect religious entities with mundane, day-to-day occurrences and

not just ritual or exceptional circumstances. The tooth fairy merits causal mention only when a tooth is lost. Santa comes up only around Christmastime. Germs could come up anytime at all: any time of day, any day of the week, any month of the year. Germs, then, are more relevant, more worth thinking about, and are not set apart as exceptional. Harris offers that the exceptionality of the tooth fairy and Santa may encourage children to think of them as not of the same status as oxygen, germs, and the like. Similarly, if you talk of gods, the ancestors, or whatever other religious entities you regard as real only during rituals or in places of worship or sacred spaces, or on special days such as Sundays or holidays, then you are tacitly communicating that these religious beings are relevant only under special conditions. Again, this marks religious talk as somehow different and easier to extract from life than talk about gravity or microwaves.

Anthropologists frequently comment that in small-scale traditional societies, no one talks about whether someone believes in the gods. Everyone does. The gods are taken for granted because they are as seamlessly a part of societal life and conversation as anything else. Applying these observations to God, parents, or teachers who want children to see God as just as real and as more important than germs might be advised to mention matter-of-factly how God matters to events in lots of different real-life, day-to-day settings and not confine God talk to a particular day of the week or holidays during the year.

In a similar vein, Pascal Boyer writes, "So my advice to religious proselytes would be to avoid bombarding people with cogent and coherent arguments for particular metaphysical claims and to provide them instead with many occasions where the claims in question can be used to produce relevant interpretations of particular situations. But religions do not need expert consultants, for they all do that anyway."[6] Boyer's advice is relevant to anyone trying to improve religious fluency or to solidify belief—their own, their children's, or others'. Instead of talking about beliefs, using beliefs to generate inferences, attitudes, and feelings in lots of different contexts encourages depth of useful belief.

Compare religious learning with language learning. When people learn their native language and acquire fluency, it is not through discussion of grammatical rules. They learn through hearing and using the language, with some occasional correction. They learn about how different words carry different connotations, evoke different emotional responses, and facilitate varieties of social encounters not based on a set of explicitly taught rules and generalizations, but through exercise and imitation. Likewise, deep and broad religious beliefs—the kind that produce long-term commitment—probably come about through a similar process of seeing them used and using them over and over to solve problems, inspire actions, and evoke emotions. I agree with Boyer's observation, but I think religious people often do require this reminder.

7. *Make religious ideas relevant for behavior.* Inspired by Harris's point about using religious teaching in mundane circumstances, I suggest linking religious ideas and commitments to concrete, everyday behaviors. Ideas that motivate behavior, which are relevant to making decisions regarding how to act, are easier to own and feel committed to than general platitudes. To illustrate, for those in religious traditions that include the

Book of Genesis as scripture, a common teaching based on the second and third chapters of Genesis is that human beings have a special responsibility for the rest of the world. Humans are said to be created with special privileges and special obligations as God's image bearers, and were placed in the Garden of Eden with instructions to care for the Garden. You might draw from these scriptural passages that people are obliged to care for the environment. If the teaching moment ends there, however, it is far less likely to take root in a young mind than if the next step in teaching is to go and clean up a littered roadside or clear dead brush that might be a fire hazard or even to invest some sweat in a backyard vegetable garden. Similarly, theological directives to care for the poor will mean more if their presentations are followed by serving at a local food pantry or soup kitchen, or working a day with a Habitat for Humanity project. Children's religious development would benefit from the opportunity to identify and empathize with actual individuals who are struggling with poverty and to care for them. Behavior-relevant applications could be something as small as ending teaching moments with little behavioral challenges. For instance, when teaching about biblical instructions to be submissive to governmental institutions, a relevant challenge for children might be to identify their least favorite teacher and consider how they will try to be more respectful to that teacher and why.

8. *Act on your own religious beliefs.* The previous suggestion will be easier to execute and more convincing if you, as a parent or teacher, actually act in ways motivated by your religious beliefs. If children cannot see that what you believe about God, spirit ancestors, or whatever else makes a discernible difference

in your life, they can hardly be expected to regard these beliefs as important. Additionally, children may doubt your commitment to them, and if you do not really believe, why should they take your teaching seriously?

My mother-in-law's parents regularly sent her and her eight brothers and sisters to church every Sunday. Though her father claimed to be Catholic, he sent the kids to a Lutheran church because it was nearer and they could walk without him. The parents never went to church or ever gave clear indications of their commitment to any part of the Christian tradition in word or deed. Not surprisingly, few of the children grew into adult paragons of Christian commitment and virtue. Their parents communicated through their inaction that Christianity was not important to life. Actions do speak louder than words.

9. *Associate religious commitments with a full range of emotions.* As developmental psychologist Chris Boyatzis rightly points out, religious belief in children is not just about their cool cognition. It is not just about what kids think, but also about what they feel.[7]

Emotional states are powerful cues for memories and ideas. Have you ever been in a sour mood and all you can think of is other problems or bad times? Or perhaps a sense of satisfaction and accomplishment set your mind racing on other occasions during which you felt similar feelings? Consider, then, that if children associate God only with negative emotions or only positive emotions, God will only be emotionally salient and evocative for some parts of their lives but not all parts. If God is associated only with suffering and sadness because God is seen as primarily a lifeline in times of distress, then

God will be invoked less frequently in times of joy. If God is associated only with upbeat happy songs, then God might be only the God of good times and conspicuously absent when troubles arise.

Similarly, parents, teachers, and caregivers might consider the emotional tenor of how religious teaching is offered. Is it presented in a threatening or intimidating manner? Do your children associate God talk with utter boredom? Repeated pairing of religious ideas with negative emotions, even inadvertently, will cause the negative associations to rub off on the religious ideas, hardly motivation to keep bringing them to mind later in life.

Many churches and religious ministries recognize the need to connect their belief system with positive feelings and so organize special fun events for children and youth such as camps and vacation Bible schools. The programmatic challenge is to make sure the fun is authentically an extension of the religious content and not a contrived addendum. Such special programs can be a good start in associating religion with positive emotions, but if day-in-and-day-out religious talk and teaching is associated only with somberness, melancholia, and guilt, the occasional fun event is likely to be insufficient. If you are inclined to point out to children the beauty of a sunset as material evidence of God's awesome majesty, do not neglect to observe that ducks evince God's sense of humor. Those who use the story of God engulfing Elijah's altar in flames (1 Kings 18) to illustrate God's power and sovereignty should not neglect stories such as when God has Balaam's donkey say, "What have I done to you to make you beat me these three times?" (Numbers 22:28) as instances of God's irony.

10. *Form a strong, secure relational attachment with your children.* Research in psychology of religion over many decades points to the importance of children's relationships with their parents in influencing how they will characterize their relationship with God. If their parents are harsh and emotionally distant, God is likely to be thought of as distant and wrathful. If children do not feel secure in their attachment to their parents, they are unlikely to feel securely attached to God.[8] Children who (1) trust their parents for material and emotional support, (2) can rely on parents to ensure their security through authoritative (but not authoritarian) guidelines, and (3) have parents who allow them room to explore life within sensible and predictable boundaries will tend to form secure attachments with their parents. These secure attachments will lead to children who want to adopt their parents' religious beliefs and regard God as someone with whom they want a relationship. In contrast, if parents are inconsistent in the structure and support they provide for their children's lives, are volatile, overcontrolling, harshly punitive, or capricious, children are less likely to gravitate toward their parents' god as a good, secure attachment figure. Indeed, children are likely to rebel from such parents and reject their religious association, particularly if parents have used their religion to justify such parenting.

A big part of religious belief formation is social identification. As I mentioned in Chapter 10, at age eleven, my daughter wanted to know what my religious beliefs were because she wanted to share them; she wanted to *be* what I was. If your child has a secure relationship with you as a parent, she will likely want to adopt your religious beliefs and practices as well. Like it or not, you become a role model in this context. It follows, then, that having a good rationale for your religious beliefs and practices and consistently acting on the implications of your religious convictions become all the more important as your children look to you for a model of whom they strive to be.

To summarize, if you want children to build on their natural dispositions to become mature adherents to a particular religious tradition, I recommend the following ten guidelines:

1. *Start teaching early.* A little investment in the early years—three to four years old—might be more valuable than a greater investment in later childhood. Children can handle more theology than parents think if presented in illustrative ways instead of through abstractions.
2. *Teach in love and humility.* Do not bully children into belief. Instead invite them to share in those most important commitments of your life. Do not be afraid to say that you do not know or even that you are not sure. Explore with your children.
3. *Teach them how to think, learn, and discern for themselves.* You will not have all the facts, and so modeling how to learn and how to separate bad ideas from good ideas may be more important than much of the particular content you want to convey.
4. *Don't use wishy-washy language.* If you believe God exists, talk as if God exists and don't hedge with "I believe . . ." If you trust God to be faithful, don't say, "I have faith in God"; say, "I trust God."

5. *Talk about God in contexts in which God's existence and action make a detectable difference.* Does your god make a difference in the real world? Point it out. If children occasionally detected divine activity, their belief will be reinforced.
6. *Use religious ideas in ordinary life.* If God is going to be real *and* relevant for children, then it would help if God is not confined to holy days and ritual settings.
7. *Make religion motivate actions.* Does believing in spirits or gods change the way you act or motivate particular values? Draw out the connections through actions.
8. *Act on your beliefs.* Children need behavioral evidence that you believe what you say. Kids are wonderful detectors of inconsistency and hypocrisy.
9. *Associate religious commitments with a full range of emotions.* If you want God to be relevant to all of children's feelings and experiences, God cannot be associated only with fun and games or only with awe and somberness.
10. *Form a secure relationship with your children.* Kids want to be like those they like. They like those they trust and around whom they feel secure and supported.

My final comment for adults wishing to encourage religious beliefs in children is this: *ultimately it is up to children to believe or not.* Children might be born believers, with strong natural dispositions toward religious thought and practice. Nevertheless, all the factors I identified that buffer against religious belief may be absent, all of the best teaching strategies may be in place, your best efforts may have been delivered with the utmost sincerity, care, and love, yet a child still grows up to be an unbeliever. People are ultimately free to make their own decisions. Personal, social, and other environmental factors may inform and constrain what people will tend to believe, but when it comes down to any given individual, people have ultimate control over what they believe and do. Parents, teachers, and caregivers should know that they cannot properly claim credit or blame if someone becomes a true believer or not. I hope this observation gives comfort: if you have done due diligence as a teacher, then rest easy. Though children may be born believers, whether they die believers is between them and God.

Notes

Introduction: On the Train to Jaipur

1. Paul L. Harris, Emma Brown, Crispin Marriott, Semantha Whittall, and Sarah Harmer, "Monsters, Ghosts and Witches: Testing the Limits of the Fantasy-Reality Distinction in Youth Children," *British Journal of Developmental Psychology* 9 (1991): 105–123; Henry M. Wellman and David Estes, "Early Understanding of Mental Entities: A Reexamination of Childhood Realism," *Child Development* 57 (1986): 910–923.
2. See Paul L. Harris, *The Work of the Imagination* (Oxford: Blackwell, 2000), especially chap. 4.
3. David Ian Miller, "Finding My Religion: Julia Sweeney Talks about How She Became an Atheist," *San Francisco Chronicle*, August 15, 2005, accessed January 14, 2011, http://articles.sfgate.com/2005-08-15/news/17384089_1_religious-los-angeles-dear-god. Incidentally, children are not more inclined to believe in things because they resemble cartoon characters. If anything, the contrary is true.
4. For more on why thinking of minds as infinitely malleable or featureless, see Steven Pinker, *The Blank Slate: The Modern Denial of Human Nature* (New York: Viking, 2002).
5. Andrew N. Meltzoff and N. Keith Moore, "Newborn Infants Imitate Adult Facial Gestures," *Child Development* 54 (1983): 702–709.

6. I am the first to concede that alternative glosses of the data to date may be reasonable and new data may change the state of the art as presented here. Nevertheless, for clarity and to put in sharp relief where the critical issues for future scientific research lie, I present a strong version of the thesis that children are naturally born believers.

One: Secret Agents Everywhere

1. Throughout I use *religion* to mean those shared beliefs and practices that are motivated by a belief that some kind of god exists. Gods may include spirits, ghosts, demons, or a benevolent creator God. For the presentation here, however, I take disembodied spirits or intentional beings with "spirit bodies" or nonbiological bodies instead of ordinary biological bodies to be gods for this discussion, and so, religious.
2. Robert N. McCauley, *Why Religion Is Natural and Science Is Not* (New York: Oxford University Press, 2011).
3. Elizabeth S. Spelke and Katherine D. Kinzler, "Core Knowledge," *Developmental Science* 10 (2007): 89–96.
4. Renee Baillargeon, Laura Kotovsky, and Amy Needham, "The Acquisition of Physical Knowledge in Infancy," in *Causal Cognition: A Multidisciplinary Debate*, ed. Dan Sperber, David Premack, and Ann James Premack (Oxford: Oxford University Press, 1995).
5. Elizabeth S. Spelke, Ann Phillips, and Amanda L. Woodward, "Infant's Knowledge of Object Motion and Human Action," in *Causal Cognition: A Multidisciplinary Debate*, ed. Dan Sperber, David Premack, and Ann James Premack (Oxford: Oxford University Press, 1995): see also W. A. Ball, "The Perception of Causality in the Infant" (paper presented at the meeting of the Society for Research in Child Development, Philadelphia, April 1973).
6. Elizabeth S. Spelke and Katherine D. Kinzler, "Core Knowledge," *Developmental Science* 10 (2007): 89–96.
7. One might want to quibble with the idea that cats or computers are genuine agents. We might treat them like agents, but perhaps in fact they are mindless machines that only respond to their environment, like billiard balls do. Then again, maybe they are just as much intentional agents as humans are. In a lecture, I once suggested that when our computers frustrate us and we yell at them, it is a sign that we misidentify them

as agents. One student responded that we do not *mis*identify them as agents because computers *are* agents that willfully and maliciously try to frustrate human users. But whether a particular thing is in fact an agent is irrelevant here. What matters is that children (and adults) do carve up the world into agents and nonagents and reason very differently about these two classes of things. On the attribution of agency to computers, see Batya Friedman, "It's the Computer's Fault: Reasoning about Computers as Human Agents," *Proceedings of the 2004 Conference on Human Factors in Computing Systems* (AMC Press, 1995); and Youngme Moon and Clifford Nass, "Are Computers Scapegoats? Attributions of Responsibility in Human-Computer Interaction," *International Journal of Human-Computer Studies* 49 (1998): 78–94.

8. Christine P. Ellsworth, Darwin W. Muir, and Sylvia M. J. Hains, "Social Competence and Person-Object Differentiation: An Analysis of the Still-Face Effect," *Developmental Psychology* 29 (1993): 63–73.
9. Michèle Molina et al., "The Animate-Inanimate Distinction in Infancy: Developing Sensitivity to Constraints on Human Actions," *Journal of Cognition and Development* 5 (2004): 399–426.
10. Elizabeth S. Spelke, Ann Phillips, and Amanda L. Woodward, "Infant's Knowledge of Object Motion and Human Action," in *Causal Cognition: A Multidisciplinary Debate*, ed. Dan Sperber, David Premack, and Ann James Premack (Oxford: Oxford University Press, 1995).
11. Michael Tomasello, *The Cultural Origins of Human Cognition* (Cambridge, MA: Harvard University Press, 1999).
12. György Gergely et al., "Taking the Intentional Stance at 12 Months of Age," *Cognition* 56 (1995): 165–193.
13. Gergely Csibra, "Goal Attribution to Inanimate Agents by 6.5-Month-Old Infants," *Cognition* 107 (2008): 705–717. See also György Gergely and Gergely Csibra, "Teleological Reasoning in Infancy: The Naive Theory of Rational Action," *Trends in Cognitive Sciences* 7 (2003): 287–292.
14. Susan Johnson, Virginia Slaughter, and Susan Carey, "Whose Gaze Will Infants Follow? The Elicitation of Gaze-Following in 12-Month-Olds," *Developmental Science* 1 (1998): 233–238.
15. Valerie Corkum and Chris Moore, "Origins of Joint Visual Attention in Infants," *Developmental Psychology* 34 (1998): 28–38.
16. Marjorie Taylor, *Imaginary Companions and the Children Who Create Them* (New York: Oxford University Press, 1999).

17. Incidentally, this was by no means a religious family and so the son's naming one of the invisible dogs Sin was regarded with a mix of amusement and perplexity.
18. J. Bradley Wigger, "Imaginary Companions, Theory of Mind, and God" (paper presented at the Cognition, Religion, and Theology Conference, Merton College, Oxford University, June 29, 2010), and "See-Through Knowing: Learning from Children and Their Invisible Friends," *Journal of Childhood and Religion* 2 (2011).
19. I do not mean to suggest that God is nothing more than an imaginary friend for these children. As Wigger explains, though attributed more Godlike knowledge than their visible friends, children treat God and invisible friends as importantly different from each other as well.
20. Philippe Rochat, Rachel Morgan, and Malinda Carpenter, "Young Infants' Sensitivity to Movement Information Specifying Social Causality," *Cognitive Development* 12 (1997): 537–561.
21. Philippe Rochat, Tricia Striano, and Rachel Morgan, "Who Is Doing What to Whom? Young Infants' Developing Sense of Social Causality in Animated Displays," *Perception* 33 (2004): 355–369.
22. Fritz Heider and Marianne Simmel, "An Experimental Study of Apparent Behavior," *American Journal of Psychology* 57 (1944): 243–249.
23. Ibid., p. 247.
24. Brian J. Scholl and Patrice D. Tremoulet, "Perceptual Causality and Animacy," *Trends in Cognitive Sciences* 4 (2000): 299–308.
25. Justin L. Barrett and Amanda Hankes Johnson, "The Role of Control in Attributing Intentional Agency to Inanimate Objects," *Journal of Cognition and Culture* 3 (2003): 208–314.
26. Stewart E. Guthrie, *Faces in the Clouds: A New Theory of Religion* (New York: Oxford University Press, 1993).

Two: Children in Search of a Purpose

1. Deborah Kelemen, "The Scope of Teleological Thinking in Preschool Children," *Cognition* 70 (1999): 241–273.
2. Ibid., p. 256.
3. Deborah Kelemen, "Why Are Rocks Pointy? Children's Preference for Teleological Explanations of the Natural World," *Developmental Psychology* 35 (1999): 1440–1453. Children had been given similar physical

explanations about bits of stuff piling up as good and legitimate explanations, but warming them up to these kinds of physical explanations did not change their preference for teleological explanations.

4. Deborah Kelemen and Cara DiYanni, "Intuitions about Origins: Purpose and Intelligent Design in Children's Reasoning about Nature," *Journal of Cognition and Development* 6 (2005): 3–31.
5. Ibid., pp. 29–31.
6. The order of these three options was counterbalanced so as not to favor one over the others.
7. George E. Newman et al., "Early Understandings of the Link between Agents and Order," *Proceedings of the National Academy of Sciences of the United States of America* 107 (2010): 17140–17145.
8. Babies in this experiment viewed both the disordering test and the ordering test, but half saw the disordering test first and half watched the ordering test first. As hypothesized, babies on average looked longer at the ordering test. A subsequent experiment ruled out the possibility that they just liked to look at order more than disorder.
9. Newman et al., "Early Understandings," p. 17141.
10. Just why our minds naturally see design and purpose in the world is a debated issue. One possibility is that such teleological reasoning is an adaptive strategy for figuring out how the world works and learning how different plants and animals—or their parts—might be used for human purposes.
11. "Lightning Hits Preacher after Call to God," *BBC News*, July 4, 2003, accessed January 14, 2011, http://news.bbc.co.uk/1/hi/world/americas/3044178.stm.
12. Jesse M. Bering and Becky D. Parker, "Children's Attributions of Intentions to an Invisible Agent," *Developmental Psychology* 42 (2006): 253–262.
13. Deborah Kelemen and Evelyn Rosset, "The Human Function Compunction: Teleological Explanation in Adults," *Cognition* 111 (2009): 138–143.
14. Krista Casler and Deborah Kelemen, "Development Continuity in Teleo-Functional Explanation: Reasoning about Nature among Romanian Romani Adults," *Journal of Cognition and Development* 9 (2008): 340–362.

15. Intelligent design theory is the idea that an intelligent being such as a god is needed to account for certain aspects of biological structures. That is, at points during evolution, an intelligent being intervened supernaturally to help evolution along.

Three: Identifying the Maker

1. Jean Piaget, *The Child's Conception of the World*, trans. Andrew Tomlinson (Paterson, NJ: Littlefield, Adams, 1960), p. 273.
2. Ibid., p. 269.
3. Ibid., p. 352.
4. Ibid., p. 354.
5. Ibid., p. 381.
6. Ibid.
7. Ibid., p. 354.
8. George E. Newman et al., "Early Understandings of the Link between Agents and Order," *Proceedings of the National Academy of Sciences of the United States of America* 107 (2010): 17140–17145.
9. Susan A. Gelman, "The Development of Induction within Natural Kind and Artifact Categories," *Cognitive Psychology* 20 (1988): 87–90.
10. Susan A. Gelman and Kathleen E. Kremer, "Understanding Natural Cause: Children's Explanations of How Objects and Their Properties Originate," *Child Development* 62 (1991): 396–414.
11. One difficulty in comparing Gelman's findings directly with Piaget's is that Piaget did not provide answer rates or any relevant statistics.
12. Olivera Petrovich, "Preschool Children's Understanding of the Dichotomy Between the Natural and the Artificial," *Psychological Reports* 84 (1999): 3–27.
13. Olivera Petrovich, "Understanding of Non-Natural Causality in Children and Adults: A Case Against Artificialism," *Psyche en Geloof* 8 (1997): 151–165.
14. Deborah Kelemen and Cara DiYanni, "Intuitions about Origins: Purpose and Intelligent Design in Children's Reasoning about Nature," *Journal of Cognition and Development* 6 (2005): 3–31.
15. E. Margaret Evans, "Cognitive and Contextual Factors in the Emergence of Diverse Belief Systems: Creation versus Evolution," *Cognitive Psychology* 42 (2001): 217–266.

16. Ibid., pp. 226–227.
17. Ibid., p. 227.
18. For evidence of young children's assumptions that animals have unchangeable, unseen essences and that (consequently) parents have children of the same species, see Frank C. Keil, *Concepts, Kinds, and Cognitive Development* (Cambridge, MA: MIT Press, 1989).
19. Piaget, *Child's Conception of the World*, pp. 378–379.
20. Ibid., p. 379.
21. For instance, see Freud's *Totem and Taboo* (London: Ark Paperback, 1983). Likewise, Piaget, *The Child's Conception of the World*, offered a form of the anthropomorphism hypothesis, noting that greater conceptual tools to think nonanthropomorphically arrive in late childhood and adolescence.
22. For instance, some threads of Hindu thought include a notion of a supergod (Brahma in classical Hinduism) construed in many of the same terms as the Abrahamic God: superknowing, superperceiving, superpowerful, and so forth.
23. Justin L. Barrett and Rebekah A. Richert, "Anthropomorphism or Preparedness? Exploring Children's God Concepts," *Review of Religious Research* 44 (2003): 300–312.

Four: The Mind of God

1. Douglas Adams, *The Hitchhiker's Guide to the Galaxy* (London: Pan Books, 1979), pp. 108–109.
2. On children's confusion between reality and beliefs see, for instance, Deborah Zaitchik, "When Representations Conflict with Reality: The Preschooler's Problem with False Beliefs and 'False' Photographs," *Cognition* 35 (1990): 41–68. For a general review of research in this area, see Henry M. Wellman, David Cross, and Julanne Watson, "Meta-Analysis of Theory of Mind Development: The Truth about False Belief," *Child Development* 72 (2001): 655–684.
3. Children seem to appreciate that Mom won't act on hidden things at an age before they understand Mom won't know about the hidden thing. So a two-year-old may "hide" something from parents, but the hiding seems to be a strategy to keep parents from acting rather than to keep them from knowing. This distinction is difficult for most adults be-

cause we so seamlessly combine consciously knowing about information with the ability to act on the information.

4. Jean Piaget, *The Child's Conception of the World*, trans. Andrew Tomlinson (Paterson, NJ: Littlefield, Adams, 1960); Ronald G. Goldman, *Religious Thinking from Childhood to Adolescence* (London: Routledge and Kegan Paul, 1964).
5. For a review of several experiments, see Justin L. Barrett and Rebekah A. Richert, "Anthropomorphism or Preparedness? Exploring Children's God Concepts," *Review of Religious Research* 44 (2003): 300–312.
6. Justin L. Barrett, Roxanne Moore Newman, and Rebekah A. Richert, "When Seeing Is Not Believing: Children's Understanding of Humans' and Non-Humans' Use of Background Knowledge in Interpreting Visual Displays," *Journal of Cognition and Culture* 3 (2003): 91–208.
7. Ibid., pp. 91–108.
8. Michael J. Chandler and David Helm, "Developmental Changes in the Contribution of Shared Experience to Social Role-Taking Competence," *International Journal of Behavioral Development* 7 (1984): 145–156.
9. Though treating the dog as more knowledgeable than they ought, the younger children did regard God as more knowing than the dog.
10. J. Bradley Wigger, "Imaginary Companions, Theory of Mind, and God" (paper presented at the Cognition, Religion, and Theology Conference, Merton College, Oxford University, June 29, 2010), and "See-Through Knowing: Learning from Children and Their Invisible Friends," *Journal of Childhood and Religion* 2 (2011).
11. Emily Reed Burdett and Justin L. Barrett, "Children's Intuitions of Memory in Divine, Human, and Non-Human Minds" (in preparation).
12. Nicola Knight et al., "Children's Attributions of Beliefs to Humans and God: Cross-Cultural Evidence," *Cognitive Science* 28 (2004): 235–243.
13. Nicola Knight, "Yukatek Maya Children's Attributions of Belief to Natural and Non-Natural Entities," *Journal of Cognition and Culture* 8 (2008): 235–243.
14. Nikos Makris and Dimitris Pnevmatikos, "Children's Understanding of Human and Super-Natural Mind," *Cognitive Development* 22 (2007): 365–375.
15. Marta Giménez-Dasí, Silvia Guerrero, and Paul L. Harris, "Intimations of Immortality and Omniscience in Early Childhood," *European Journal of Developmental Psychology* 2 (2005): 285–297.

16. Perhaps, then, the Greek children in Makris and Pnevmatikos's study were unfamiliar with God's omniscience or otherwise were confused about who God was. It is not uncommon for young children to confuse "God" with Jesus or even with the parish priest if parents and educators aren't careful. In a place in which people use icons of a very human-like God, it would not be surprising if children were confused.

Five: The Nature of God

1. Harvey Whitehouse, *Inside the Cult: Religious Innovation and Transmission in Papua New Guinea* (Oxford: Oxford University Press, 1995).
2. I thank anthropologist Richard Sosis for this example. He writes: "It is likely that the celebration of Rosh Chodesh did not mix well with the ominous feeling of Rosh Hashana and the 10 days of repentance (or Days of Awe as they are known) which follow leading up to Yom Kippur, hence the delay in blessing the moon. That's the pragmatic explanation. A more folksy explanation is that we delay from blessing the new moon in Tishrei to fool the Satan so that he does not know the exact date of Yom Kippur (this explanation is found in Rabbi Mordecai Jaffe's 'Levush' [literally 'garment'] written in Poland in the 16th century. Levush is a fairly obscure rabbinic code but this explanation is not uncommon") (personal communication, March 2008).
3. I do not mean this observation as a sleight. Many ideas make little sense to people but may be true, nevertheless. The sciences are full of examples.
4. John H. Flavell et al., "Young Children's Understanding of Fact Beliefs versus Value Beliefs," *Child Development* 61 (1990): 915–928.
5. Justin L. Barrett, Rebekah A. Richert, and Amanda Driesenga, "God's Beliefs versus Mother's: The Development of Nonhuman Agent Concepts," *Child Development* 72 (2001): 58–60.
6. Nikos Makris and Dimitris Pnevmatikos, "Children's Understanding of Human and Super-Natural Mind," *Cognitive Development* 22 (2007): 365–375.
7. Rebekah A. Richert and Justin L. Barrett, "Do You See What I See? Young Children's Assumptions about God's Perceptual Abilities," *International Journal for the Psychology of Religion* 15 (2005): 283–295.

8. Marta Giménez-Dasí, Silvia Guerrero, and Paul L. Harris, "Intimations of Immortality and Omniscience in Early Childhood," *European Journal of Developmental Psychology* 2 (2005): 288. The questions were:

   > "Right now there aren't any dinosaurs in the world. But a long time ago there were lots of dinosaurs in the world, like this (show picture). Now what about ___? When there were dinosaurs in the world, did ___ exist?"
   >
   > "Right now—you're a little boy/girl but a long time ago you were a little baby right? How about—___? Was s/he a little baby a long time ago?"
   >
   > "What's going to happen to ___ next year and the year after that? Will he get older and older or will he stay just the same?"
   >
   > "What will happen to—___ a long, long time from now? Will ___ die or will s/he go on living for ever and ever?"

9. Statistical analyses and comments from parents showed that the question about the dinosaur created some confusion and was measuring something different from the other three items, so we dropped it from our analyses.
10. A common explanation for why it is that people around the world and throughout history have tended to believe in some kind of afterlife is wish fulfillment. People are naturally afraid of death (because fearing death is good for survival), and this fear of death leads us to search for a more comforting alternative to soothe our anxiety. Hence, people spontaneously invent the idea of an afterlife, which makes them less fearful of death. I hope the problems with such an account are immediately obvious. If our human nature has endowed us with a fear of death to help us survive and reproduce, being able to wish it away would be a severe disability. We would expect non–death fearers to cavalierly charge into calamity and quickly be eliminated from the gene pool. It wouldn't take long for only afterlife skeptics to be around. Likewise, many conceptions of the afterlife are none too comforting or attractive. The Greeks had the grim place of the dead on the other side of the River Styx. Ancient Hebrews had Sheol, a shadowy, marginal existence that King David begged for God to rescue him from. Ancient Egyptians pictured a weighing of the heart—the seat of your identity—followed either by a passing into the next life bearing many similarities to the

present one (if you were virtuous enough) or by the crocodile-headed monster Ammit eating the heart. None of these among numerous other afterlives represent an idyllic retirement community that relieves anxiety over death. On the contrary, the modern, secular view that you simply stop existing should carry the least fear if coolly considered. If wish fulfillment were the answer to why people believe what they do about death, some kind of utter and complete extinction of the self at death would be a stronger candidate than many of the religious ideas out there, and yet afterlife ideas persist.

11. For instance, see Jesse M. Bering, "Intuitive Conceptions of Dead Agents' Minds: The Natural Foundations of Afterlife Beliefs as Phenomenological Boundary," *Journal of Cognition and Culture* 2 (2002): 263–308; Paul Bloom, *Descartes' Baby: How the Science of Child Developmen Explains What Makes Us Human* (London: Heinemann, 2004); Jesse M. Bering, "The Folk Psychology of Souls," *Behavioral and Brain Sciences* 29 (2006): 453–462; Rita Astuti and Paul L. Harris, "Understanding Mortality and the Life of the Ancestors in Rural Madagascar," *Cognitive Science* 32 (2008): 713–740.
12. The afterlife beliefs of Christians in these cases are instances of what D. Jason Slone, *Theological Incorrectness: Why Religious People Believe What They Shouldn't* (New York: Oxford University Press, 2004), termed *theological incorrectness*—a tendency to believe (often unknowingly) distorted versions of theological positions because they are more natural, a theme I develop more in Chapter 6.
13. H. Clark Barrett, "Human Cognitive Adaptations to Predators and Prey" (Ph.D. diss., University of California, Santa Barbara, 1999): Jesse M. Bering and David F. Bjorklund, "The Natural Emergence of Reasoning about the Afterlife as a Developmental Regularity," *Developmental Psychology* 40 (2004): 217–233; H. Clark Barrett and Tanya Behne, "Children's Understanding of Death as the Cessation of Agency: A Test Using Sleep versus Death," *Cognition* 96 (2005): 93–108; Paul L. Harris and Marta Giménez, "Children's Acceptance of Conflicting Testimony: The Case of Death," *Journal of Cognition and Culture* 5 (2005): 143–164; Astuti and Harris, "Understanding Mortality."
14. Jesse M. Bering, Carlos Hernández-Blasi, and David F. Bjorklund, "The Development of Afterlife Beliefs in Religiously and Secularly Schooled Children," *Developmental Psychology* 23 (2005): 587–607.

15. Jesse Bering, Paul Bloom, and Pascal Boyer all offer accounts along these lines but with differing emphases and possible mechanisms. Bering's perspective recognizes the dissociation between minds and bodies but regards the key difference being how we can successfully imagine the termination of mental versus biological activities. Bering has argued that at least one reason that we find belief in the afterlife so intuitive is that we have a hard time imagining the complete end to our own mental states. That is, we cannot simulate to ourselves what it would be like to not be aware, think, or know. As Robert Hinde observed, "It is difficult to imagine nonexistence, because one is imagining oneself as unable to imagine." Because we cannot imagine what it would be like to cease existing at death, we find suggestions that spirits live on very intuitively satisfying, argues Bering. It does not appear to me that Bering, Bloom, and Boyer's accounts are entirely incommensurable. Robert A. Hinde, *Why Gods Persist: A Scientific Approach to Religion* (London: Routledge, 1999); Pascal Boyer, *Religion Explained: The Evolutionary Origins of Religious Thought* (New York: Basic Books, 2001); Bloom, *Descartes' Baby*; Jesse M. Bering, "The Folk Psychology of Souls," *Behavioral and Brain Sciences* 29 (2006): 453–462.
16. Leonard D. Katz, ed., *Evolutionary Origins of Morality: Cross-Disciplinary Perspectives* (Thoverton, UK: Imprint Academic, 2000); Marc D. Hauser, *Moral Minds: How Nature Designed Our Universal Sense of Right and Wrong* (New York: Ecco/HarperCollins, 2006); Jonathan Haidt, "The New Synthesis in Moral Psychology," *Science* 316 (2007): 998–1002.
17. Hauser, *Moral Minds.*
18. For instance, see Richard Dawkins, *The God Delusion* (London: Bantam Press, 2006); Daniel C. Dennett, *Breaking the Spell: Religion as a Natural Phenomenon* (London: Allen Lane, 2006).
19. The metaphysical concept of the Tao originates in Taoism and features in Buddhism, Confucianism, and other religious and philosophical systems.
20. C. S. Lewis, *Mere Christianity* (New York: Macmillan, 1960), p. 17.
21. Boyer, *Religion Explained*, p. 189 (emphasis in original).
22. Ibid.
23. Philosopher Richard Swinburne offers a more developed but related argument in *Is There a God?* (Oxford: Oxford University Press, 1996), and briefly synopsized in *Was Jesus God?* (Oxford: Oxford University

Press, 2008). He argues that if God is perfectly free to actualize any circumstances (a component of being all-powerful), and God is all-knowing, then God knows the best way to act and, hence, is perfectly good.

24. John H. Flavell, Patricia H. Miller, and Scott A. Miller, *Cognitive Development* (Englewood Cliffs, NJ: Prentice Hall, 1993).
25. For more on language acquisition, see ibid.; Steven Pinker, *The Language Instinct: How the Mind Creates Language* (New York: HarperPerennial, 1995).
26. Peter K. Gregersen et al., "Early Childhood Music Education and Predisposition to Absolute Pitch: Teasing Apart Genes and Environment," *American Journal of Medical Genetics* 98 (2001): 280–282.
27. Jean Piaget, *The Child's Conception of the World*, trans. Andrew Tomlinson (Paterson, NJ: Littlefield, Adams, 1960), p. 378.
28. I thank Emma Burdett for sharing this account with me. Note, too, that in this case, the five-year-old was correct, not necessarily about the existence of God but about the correct answers. I may not believe in unicorns, but if you asked me whether they have horns, the correct answer is surely "Yes," not "No, because they don't exist."

Six: Natural Religion

1. Incidentally, this disagreement illustrates another way in which theology (but not religion) and science have similarities. Science occasionally makes discoveries about nature that seem mutually contradictory and absolutely true at the same time. Critics condemn the scientific enterprise in question as obviously flawed, but eventually solutions to the mysterious contradiction are discovered—sometimes a solution that remains impossible to fully understand but is nonetheless true.
2. In my book *Why Would Anyone Believe in God?* (Walnut Creek, CA: AltaMira Press, 2004), I argued that some properties of God from the great monotheisms enjoy a cognitive advantage over lesser gods. I did not mean to imply there or here that strict monotheism (believing in the existence of only one God and no saints, ghosts, spirits, devils, or any other supernatural agents) was the most cognitively natural position.
3. In *Religion Explained: The Evolutionary Origins of Religious Thought* (New York: Basic Books, 2001), anthropologist Pascal Boyer has suggested that across religious traditions, gods generally are regarded only as pay-

ing attention to those sorts of things that humans really care about, especially the topics of gossip: who does what with or to whom. He even suggests that with all-knowing gods (as in Christianity, Islam, and Judaism), people think of them as all-knowing only in an abstract sense, but in real time they represent these gods as knowing things like whether I have been naughty or nice and not trivia such as the number of bacteria living in my intestines.

4. Other devotional religions have similar notions about throwing oneself on the mercy of the divine, but I focus on Christianity here because I know it best.
5. Donald McCullough, *If Grace Is So Amazing, Why Don't We Like It?* (San Francisco: Jossey-Bass, 2005), 4–5 (emphasis in original).
6. Leda Cosmides and John Tooby, "Evolutionary Psychology and the Generation of Culture, Part II: Case Study: A Computational Theory of Social Exchange," *Ethology and Sociobiology* 10 (1989): 51–97.
7. For a skillful and helpful discussion of gratitude, see Robert A. Emmons, *Thanks! How Practicing Gratitude Can Make You Happier* (New York: Houghton Mifflin, 2008). See also Robert A. Emmons and Michael E. McCullough, eds., *The Psychology of Gratitude* (New York: Oxford University Press, 2004).
8. I thank Nick Gibson for his perspectives on how cognitively natural or unnatural grace might be.
9. Matthew 19:14, NIV.
10. For evidence of early developing distinctions between animate things and inanimate things, see Elizabeth S. Spelke, Ann Phillips, and Amanda L. Woodward, "Infants' Knowledge of Object Motion and Human Action," in *Casual Cognition: A Multidisciplinary Debate*, ed. Dan Sperber, David Premack, and Ann James Premack (New York: Oxford University Press, 1995), 44–78; and other essays in Dan Sperber, David Premack, and Ann James Premack, eds., *Causal Cognition: A Multidisciplinary Debate* (New York: Oxford University Press, 1995).
11. On transforming animals from one kind into another see Frank C. Keil, *Concepts, Kinds, and Cognitive Development* (Cambridge, MA: MIT Press, 1989); on insides of animals, see Daniel J. Simons and Frank C. Keil, "An Abstract to Concrete Shift in the Development of Biological Thought: The *Insides* Story," *Cognition* 56 (1995): 129–163. For a more recent general review of the cross-cultural scientific evidence,

see Kayoko Inagaki and Giyoo Hatano, *Young Children's Naive Thinking about the Biological World* (New York: Psychology Press, 2002).

12. For a discussion of the evidence, see Henry M. Wellman and Susan A. Gelman, "Knowledge Acquisition in Foundational Domains," in vol. 2 of *Handbook of Child Psychology,* ed. William Damon (Hoboken, NJ: Wiley, 1998), 523–573.
13. Incidentally, for the reason of these differences between science versus folk knowledge and theology versus religion, philosopher of science Robert McCauley observes that much of the discussion regarding the nature and relationship between science and religion is often making the wrong comparisons. Science is to theology as folk knowledge is to religion. See McCauley, *Why Religion Is Natural and Science Is Not* (New York: Oxford University Press, 2011).
14. Harvey Whitehouse, "Apparitions, Orations, and Rings: Experience of Spirits in Dadul" in *Spirits in Culture, History, and Mind*, ed. Jeannette Mageo and Alan Howard (New York: Routledge, 1996), p. 175.
15. Ibid., p. 176.
16. Mohammad Zia Ullah, *Islamic Concept of God* (London: Kegan Paul, 1984), p. 19.
17. Gordon Spykman, *Reformational Theology: A New Paradigm for Doing Dogmatics* (Grand Rapids, MI: Wm. B. Eerdmans, 1992), pp. 64–65.
18. Emma Cohen, *The Mind Possessed: The Cognition of Spirit Possession in an Afro-Brazilian Religious Tradition* (New York: Oxford University Press, 2007), p. 107.
19. Ibid., p. 111.
20. For details of these experiments and more thorough discussion of the rationale and implications, see Justin L. Barrett and Frank C. Keil, "Conceptualizing a Nonnatural Entity: Anthropomorphism in God Concepts," *Cognitive Psychology* 31 (1996): 219–247; Justin L. Barrett and Brant VanOrman, "The Effects of Image-Use on God Concepts," *Journal of Psychology and Christianity* 15 (1996): 38–45; Justin L. Barrett, "Theological Correctness: Cognitive Constraint and the Study of Religion," *Method and Theory in the Study of Religion* 11 (1999): 325–339.
21. Frederic C. Bartlett, *Remembering: A Study in Experimental and Social Psychology* (Cambridge: Cambridge University Press, 1995).
22. Justin L. Barrett and Melanie A. Nyhof, "Spreading Non-Natural Concepts: The Role of Intuitive Conceptual Structures in Memory and

Transmission of Cultural Materials," *Journal of Cognition and Culture* 1 (2001): 69–100.

23. Marcia K. Johnson, John D. Bransford, and Susan K. Solomon, "Memory for Tacit Implications of Sentences," *Journal of Experimental Psychology* 98 (1973): 203–205. For explanation and elaboration, see Daniel Reisberg, *Cognition: Exploring the Science of the Mind* (New York: Norton, 1997).
24. Barrett and Keil, "Conceptualizing a Nonnatural Entity, Anthropomorphism in God Concepts," *Cognitive Psychology* 31 (1996): Barrett and VanOrman, "The Effects of Image-Use on God Concepts."
25. Barrett and Keil, "Conceptualizing a Nonnatural Entity," p. 224.
26. For details, see Justin L. Barrett, "Cognitive Constraints on Hindu Concepts of the Divine," *Journal for the Scientific Study of Religion* 37 (1998): 608–619.
27. Travis Chilcott and Raymond F. Paloutzian, "The Cultivation of Divine Intimacy and Its Relation to Anthropomorphic Attribution: An Experimental Ethnographic Study on the Cognitive Effects of Gaudiya Vaishnava Religious Practices and Beliefs" (paper presented at the Cognition, Religion, and Theology Conference, Oxford University, June 28–July 1, 2010).

Seven: It's Okay to Be Childish

1. Sigmund Freud, *The Future of an Illusion*, trans. James B. Strachey (London: Norton, 1989), 42.
2. Ibid., p. 32.
3. Ibid., p. 30.
4. Adrienne Burke, "The God Delusion: Richard Dawkins," *New York Academy of Science, Science and the City Podcasts*, October 6, 2006, accessed January 13, 2011. http:www.nyas.org/Publications/Media/Podcast Detail.aspx?cid=a4bb550a-82b2-4a95-8ce8-a495c35ac0c0.
5. Alister McGrath, *Dawkins' God: Genes, Memes, and the Meaning of Life* (Oxford: Blackwell, 2005).
6. Matthew 19:14, NIV.
7. Matthew 18:3–4, NIV.
8. Gilbert K. Chesterton, *The Everlasting Man* (New York: Image Books, 1955), p. 16 (see also pp. 29–35).

Eight: So Stupid They'll Believe Anything?

1. Anthony C. Grayling, "Onward Christian Teachers?" *Guardian*, November 12, 2007, accessed January 13, 2010. http://commentisfree.guardian.co.uk/ac_grayling/2007/11/onward_christian_teachers.html.
2. For a classic example, see Edward E. Evans-Pritchard's description of the Azande in *Witchcraft, Oracles and Magic among the Azande* (Oxford: Clarendon Press, 1976).
3. Christopher Hitchens, *God Is Not Great: How Religion Poisons Everything* (New York: Twelve, 2007), pp. 219–220.
4. This assumes that religious and nonreligious people comarry without discrimination and parents have only two children each. A population of 100,000 religious people and zero nonreligious would be only approximately 90 percent religious in the next generation. After two generations, this population would be approximately 73 percent religious, and then only around 48 percent religious after the third generation, 21 percent after the fourth, and 4 percent after the fifth. If it takes two religiously indoctrinating parents to hoodwink kids at a 90 percent success rate, before long there would not be many kids who have two religious parents, and religious ideas would die out quickly.
5. I have heard these sorts of observations used to argue against religious belief. The argument often goes like this: *You believe in God only because of how you were raised, but if you grew up in another time or another place, you would believe in some other god, so you should not believe in God.* This reasoning is shoddy in a number of ways. One easy way to see the problem is to play the argument the other way around: *You believe in no God because of how you were raised, but if you were raised in another time or another place, you would believe in some god, so you should not believe in no God.* This kind of argument makes the mistake of confusing the *causes* for beliefs with *reasons* for beliefs. Another way to make the problem with the argument obvious is to reframe it. A modern city dweller from Oxford, England, decides to go for a walk in the Amazon. He is accompanied by a tribal hunter from Africa, a hunter-gatherer from Asia, and a backwoods outdoorsman from North America. On their walk, deep in the Amazon rain forest, they come across the fresh remains of an animal, they see scratch marks on some trees, and they hear a low rumbling sound. The African says, "There is a lion nearby so we should leave." The Asian

says, "There is a tiger nearby so we should leave." The North American says, "There is a bear nearby so we should leave." The Oxfordian replies to his fellow trekkers, "Look. You three can't seem to agree. It is clear that if you, my African friend, had been born in Asia, you would believe there was a tiger nearby. If you, my Asian friend, had been born in North America, you would believe there is a bear nearby. Clearly, the rational thing is not to believe any of you. There is nothing nearby and your desire to flee is merely the product of your particular up-bringing." Guess who was eaten later that day?

6. Richard Dawkins, *The God Delusion* (London: Bantam Press, 2006).
7. Daniel C. Dennett, *Breaking the Spell: Religion as a Natural Phenomenon* (London: Allen Lane, 2006); Hitchens, *God Is Not Great.* Dennett takes up the idea that theistic beliefs are well-adapted memes that shield themselves from rational or scientific scrutiny—the "spell" that he wants to see broken. He does not, however, suggest that children are susceptible to any and all ideas their parents might propose. Dennett knows cognitive science too well to make that mistake.
8. Dawkins, *God Delusion*, p. 176.
9. Ibid., p. 174.
10. Ibid., p. 176.
11. The idea that children are utterly and hopelessly gullible is not new. Eighteenth-century philosopher Thomas Reid wrote that what he calls the "credulity disposition" is "unlimited in children." Reid, however, regarded the credulity disposition—a tendency to believe the testimony of others—as perfectly rational under normal circumstances. See Thomas Reid, *Inquiry and Essays*, ed. Ronald E. Beanblossom and Keith Lehrer (Indianapolis, IN: Hackett, 1983), p. 95. See also Thomas Reid's *Inquiry into the Human Mind on the Principles of Common Sense*, ed. Derek R. Brookes (Edinburgh: Edinburgh University Press, 1997).
12. I have in mind David G. Myers's excellent *Psychology* (New York: Freeman, 2009).
13. Dawkins, *God Delusion,* p. 176.
14. Marc D. Hauser, *Moral Minds: How Nature Designed Our Universal Sense of Right and Wrong* (New York: Ecco/HarperCollins, 2006).
15. Scott Atran, *In Gods We Trust: The Evolutionary Landscape of Religion* (New York: Oxford University Press, 2002), p. 57.

16. E. Margaret Evans, "Cognitive and Contextual Factors in the Emergence of Diverse Belief Systems: Creation versus Evolution," *Cognitive Psychology* 42 (2001).
17. Joseph Henrich and Robert Boyd, "The Evolution of Conformist Transmission and Between Group Differences," *Evolution and Human Behavior* 19 (1998): 215–241; Joseph Henrich and Francisco Gil-White, "The Evolution of Prestige: Freely Conferred Status as a Mechanism for Enhancing the Benefits of Cultural Transmission," *Evolution and Human Behavior* 22 (2001): 165–196.
18. Nicholas Humphrey, *The Mind Made Flesh: Frontiers of Psychology and Evolution* (Oxford: Oxford University Press, 2002), 317.
19. See Dawkins, *God Delusion*, 191–201, where he discusses religious ideas in terms of memes and memeplexes. One concern I have about meme theory and memetics, at least as presented with regard to beliefs in gods by Dawkins, and also by philosopher Dan Dennett in *Breaking the Spell: Religion as a Natural Phenomenon* (London: Allen Lane, 2006), is that readers will come away with the idea that being a catchy idea is all about an idea protecting itself from scrutiny, as if ideas are intentional beings that can protect themselves. I trust that Dawkins and Dennett would not want their readers to misunderstand them in this way. The real action, in terms of what people believe or not, is not on the idea side of things (as the idea-as-virus metaphor might imply) but on the mind side of things. What is it about how human minds are put together and the biases they have that make them likely to generate and communicate some ideas more readily than others? After all, ideas do not just float around out there waiting to attack unsuspecting human heads. The idea-as-virus metaphor makes it seem as though ideas reproduce themselves—indeed, "good memes" are often talked about as good self-replicators. The reality is that ideas cannot do anything, let alone self-replicate. Human minds do the work. Minds construct certain thoughts, ideas, and beliefs. In *Explaining Culture: A Naturalistic Approach* (Oxford: Blackwell, 1996), anthropologist Dan Sperber has offered a more fruitful approach for explaining why some ideas spread better than others. Unfortunately, he does not have a catchy term like *memetics* to label his approach, which is more cumbersomely termed *epidemiology of representations*.

Nine: Is Atheism Unnatural?

1. Amos Tversky and Daniel Kahneman, "Judgments under Uncertainty: Heuristics and Biases," in *Judgment and Decision Making: An Interdisciplinary Reader*, ed. Terry Connoly, H. R. Arkes, and K. R. Hammond (Cambridge: Cambridge University Press, 2000), pp. 35–52.
2. Donn E. Byrne, *The Attraction Paradigm* (New York: Academic Press, 1971).
3. Yao Xinzhong and Paul Badham, *Religious Experience in Contemporary China* (Cardiff: University of Wales Press, 2007).
4. See William S. Bainbridge, "Atheism," *Interdisciplinary Journal of Research on Religion* 1 (2005): art. 2, for a discussion of atheism and figures from one nonrandom international sample and another more representative American sample.
5. Eugene Winograd and Ulric Neisser, eds., *Affect and Accuracy in Recall: Studies of "Flashbulb" Memories* (Cambridge: Cambridge University Press, 1992).
6. Simon Baron-Cohen, "The Extreme Male Brain Theory of Autism," *Trends in Cognitive Sciences* 6 (2002): 248–254.
7. Simon Baron-Cohen and Sally Wheelwright, "The Empathy Quotient (EQ): An Investigation of Adults with Asperger Syndrome or High Functioning Autism, and Normal Sex Differences," *Journal of Autism and Developmental Disorders* 34 (2004): 163–175; Rebecca C. Knickmeyer and Simon Baron-Cohen, "Foetal Testosterone and Sex Differences in Typical Social Development and in Autism," *Journal of Child Neurology* 48 (2006): 825–845.
8. See Bainbridge, "Atheism"; Benjamin Beit-Hallahmi, "Atheists: A Psychological Profile," in *The Cambridge Companion to Atheism*, ed. Michael Martin (Cambridge: Cambridge University Press, 2007), pp. 300–317.
9. Raymond F. Paloutzian, *Invitation to the Psychology of Religion* (Needham Heights, MA: Allyn & Bacon, 1996).
10. Richard Dawkins, *The God Delusion* (London: Bantam Press, 2006).
11. David G. Myers, *A Friendly Letter to Skeptics and Atheists: Musings on Why God Is Good and Faith Isn't Evil* (San Francisco: Jossey-Bass, 2008), 22–25.
12. Benson Saler and Charles A. Ziegler, "Atheism and the Apotheosis of Agency," *Temenos* 42 (2006): 7–41.

13. William S. Bainbridge, "Atheism," *Interdisciplinary Journal of Research on Religion* 1 (2005): 7.
14. Bainbridge interprets this as both a difficulty maintaining atheism while being a parent, but of course it may be that atheists are less interested in having kids in the first place.
15. Stewart E. Guthrie, *Faces in the Clouds: A New Theory of Religion* (Oxford: Oxford University Press, 1993).
16. On how existential security in contemporary western and northern Europe may prove fertile ground for nonbelief in gods, see Jonathan A. Lanman, "A Secular Mind: Towards a Cognitive Anthropology of Atheism" (Ph.D. diss., University of Oxford, 2009).
17. John D. Barrow, Frank J. Tipler, and John A. Wheeler, *The Anthropic Cosmological Principle* (Oxford: Oxford University Press, 1988); Simon Conway Morris, *Life's Solution: Inevitable Humans in a Lonely Universe* (Cambridge: Cambridge University Press, 2003); Antony Flew, *There Is a God: How the World's Most Notorious Atheist Changed His Mind* (New York: Harper, 2007).
18. In *The Blind Watchmaker: Why the Evidence of Evolution Reveals a Universe without Design* (New York: Norton, 1986), Richard Dawkins candidly confesses how difficult it is even for experts on evolution and natural selection to avoid this kind of anthropomorphic language.

Ten: Should You Introduce Children to God?

1. Christopher Hitchens, *God Is Not Great: How Religion Poisons Everything* (New York: Twelve, 2007), p. 218.
2. Ibid., p. 220.
3. Nicholas Humphrey, *The Mind Made Flesh: Frontiers of Psychology and Evolution* (Oxford: Oxford University Press, 2002), p. 291.
4. Richard Dawkins, *The God Delusion* (London: Bantam Press, 2006), p. 317.
5. Ibid., p. 315.
6. Ibid., p. 318.
7. Kathleen A. Kendall-Tackett, Linda Meyer Williams, and David Finkelhor, "Impact of Sexual Abuse on Children: A Review and Synthesis of Recent Empirical Studies," in *Children and the Law: The Essential Readings*, ed. Ray Bull (Oxford: Blackwell, 2001), pp. 31–70.

8. Matthew 5:22, NAS.
9. Exodus 20:7, NAS.
10. Dawkins, *God Delusion*, p. 327 (emphasis in original).
11. Ibid. (emphasis in original).
12. Nicholas Humphrey, *The Mind Made Flesh: Frontiers of Psychology and Evolution* (Oxford: Oxford University Press, 2002), 313.
13. One answer to this question that I would have sympathy toward is that "religious" ideas are part of a tradition of knowing that should be granted some deference over individual whims. Indeed, it is the height of intellectual arrogance to cavalierly dismiss an entire belief system (religious or otherwise) that is built on the collective intellectual efforts of thousands or millions of others without giving the belief system careful consideration. Humility and reason demand that we give intellectual traditions certain weight, if only because there is plenty of evidence that the intellectual efforts of thousands challenging and refining each other generally outpace the accomplishments of any individual. Note, however, that being "religious" does not mean reasons do not apply and merely being "religious" is not enough to merit initial deference. In high school I knew a girl who claimed her "religious" belief was that after death, if you led a basically moral life, you became a star. Though "religious" in a certain respect, such a belief was the product not of generations of scrutiny and reflection but of idiosyncratic teenage whimsy. Likewise, a tradition of spurning reason in a religious belief system does not provide grounds for giving the belief system a certain benefit of the doubt.
14. Roger Trigg, *Religion in Public Life: Must Faith Be Privatized?* (Oxford: Oxford University Press, 2007), pp. 66–67.
15. Ibid., p. 58.
16. For reviews of scientific research in this area, see Kenneth I. Pargament, *The Psychology of Religion and Coping: Theory, Research, Practice* (London: Guilford Press, 1997); Robert A. Emmons, *The Psychology of Ultimate Concerns: Motivation and Spirituality in Personality* (London: Guilford Press, 1999). A limitation on this research is that the bulk concerns religious believers in western Europe and North America, and consequently primarily considered Christians of various sorts.
17. Robert A. Emmons, *The Psychology of Ultimate Concerns: Motivation and Spirituality in Personality* (London: Guilford Press, 1999).

Eleven: Encouraging Children's Religious Development

1. A mentor of mine, a former professor whom I wanted to be like in many respects, once commented near the end of our formal relationship that he never encouraged me to become an academic like him because he was uncomfortable with making others in his own image. That job is reserved for God.
2. Roger Trigg, *Religion in Public Life: Must Faith Be Privatized?* (Oxford: Oxford University Press, 2007).
3. Apparent inconsistencies are red flags but not deal breakers when it comes to any set of ideas. Two ideas may appear to be in contradiction (say, perfect divine knowledge of the future and human free will) but may not be contradictory in reality. Often failures to see how two or more ideas can be reconciled indicate a lack of effort or creativity rather than actual incoherence in the set of ideas.
4. Paul L. Harris and Melissa A. Koenig, "Trust in Testimony: How Children Learn about Science and Religion," *Child Development* 77 (2006): 505–524.
5. I recognize I am glossing over some complex questions from the philosophy of science here and that in some cases the emergence of good scientific explanations does present challenges for religious explanations. These occasions are, however, much rarer than many science-educated people suppose. I take up some of these compatibility issues in my book *Cognitive Science, Religion, and Theology* (West Conshohocken, PA: Templeton Press, 2011). See also Malcolm Jeeves and Warren S. Brown, *Neuroscience, Psychology, and Religion: Illusions, Delusions, and Realities about Human Nature* (West Conshohocken, PA: Templeton Press, 2009). Some readers may object that multiple explanations of this sort violate Ockham's razor—that we should not unnecessarily multiply explanations. This principle, however, applies not to these cases I am pointing to here but to multiple explanations of the same kind that explain precisely the same thing. Early advocates of this heuristic for science such as Isaac Newton and William of Ockham himself did not see scientific explanations as conflicting with religious ones as both were devout Christians.
6. Pascal Boyer, *Religion Explained: The Evolutionary Origins of Religious Thought* (New York: Basic Books, 2001), p. 317.

7. Chris J. Boyatzis, "Religious and Spiritual Development in Childhood," in *Handbook of the Psychology of Religion and Spirituality*, ed. Raymond F. Paloutzian and Crystal L. Park (New York: Guilford Press, 2005), pp. 123–143.
8. Pehr Granqvist and Lee A. Kirkpatrick, "Religious Conversion and Perceived Childhood Attachment: A Meta-Analysis," *International Journal for the Psychology of Religion* 14 (2004): 223–250; Pehr Granqvist, "Building a Bridge between Attachment and Religious Coping: Tests of Moderators and Mediators," *Mental Health, Religion, and Culture* 8 (2005): 35–47; Lee A. Kirkpatrick, *Attachment, Evolution, and the Psychology of Religion* (New York: Guilford Press, 2005).

References

Adams, Douglas. 1979. *The Hitchhiker's Guide to the Galaxy*. London: Pan Books.

Astuti, Rita, and Paul L. Harris. 2008. Understanding Mortality and the Life of the Ancestors in Rural Madagascar. *Cognitive Science* 32:713–740.

Atran, Scott. 2002. *In Gods We Trust: The Evolutionary Landscape of Religion*. Oxford: Oxford University Press.

Baillargeon, Renee, Laura Kotovsky, and Amy Needham. 1995. The Acquisition of Physical Knowledge in Infancy. In *Causal Cognition: A Multidisciplinary Debate*, edited by D. Sperber, D. Premack, and A. J. Premack. Oxford: Oxford University Press.

Bainbridge, William S. 2005. Atheism. *Interdisciplinary Journal of Research on Religion* 1: art. 2.

Ball, W. A. 1973. The Perception of Causality in the Infant. Paper presented to the Society for Research in Child Development, Philadelphia, PA.

Baron-Cohen, Simon. 2002. The Extreme Male Brain Theory of Autism. *Trends in Cognitive Sciences* 6:248–254.

Baron-Cohen, Simon, and Sally Wheelwright. 2004. The Empathy Quotient (EQ): An Investigation of Adults with Asperger Syndrome or High Functioning Autism, and Normal Sex Differences. *Journal of Autism and Developmental Disorders* 34:163–175.

Barrett, H. Clark. 1999. *Human Cognitive Adaptations to Predators and Prey*. Santa Barbara: University of California, Santa Barbara.

Barrett, H. Clark, and Tanya Behne. 2005. Children's Understanding of Death as the Cessation of Agency: A Test Using Sleep versus Death. *Cognition* 96:93–108.

Barrett, Justin L. 1998. Cognitive Constraints on Hindu Concepts of the Divine. *Journal for the Scientific Study of Religion* 37:608–619.

———. 1999. Theological Correctness: Cognitive Constraint and the Study of Religion. *Method and Theory in the Study of Religion* 11:325–339.

———. 2004. *Why Would Anyone Believe in God?* Walnut Creek, CA: AltaMira Press.

———. 2011. *Cognitive Science, Religion, and Theology.* West Conshohocken, PA: Templeton Press.

Barrett, Justin L., and Amanda Hankes Johnson. 2003. The Role of Control in Attributing Intentional Agency to Inanimate Objects. *Journal of Cognition and Culture* 3:208–314.

Barrett, Justin L., and Frank C. Keil. 1996. Conceptualizing a Non-Natural Entity: Anthropomorphism and God Concepts. *Cognitive Psychology* 31:219–247.

Barrett, Justin L., Roxanne Moore Newman, and Rebekah A. Richert. 2003. When Seeing Does Not Lead to Believing: Children's Understanding of the Importance of Background Knowledge for Interpreting Visual Displays. *Journal of Cognition and Culture* 3:91–108.

Barrett, Justin L., and Melanie Nyhof. 2001. Spreading Non-Natural Concepts: The Role of Intuitive Conceptual Structures in Memory and Transmission of Cultural Materials. *Journal of Cognition and Culture* 1:69–100.

Barrett, Justin L., and Rebekah A. Richert. 2003. Anthropomorphism or Preparedness? Exploring Children's God Concepts. *Review of Religious Research* 44:300–312.

Barrett, Justin L., Rebekah A. Richert, and Amanda Driesenga. 2001. God's Beliefs versus Mom's: The Development of Natural and Non-Natural Agent Concepts. *Child Development* 72:50–65.

Barrett, Justin L., and Brant VanOrman. 1996. The Effects of Image Use in Worship on God Concepts. *Journal of Psychology and Christianity* 15:38–45.

Barrow, John D., Frank J. Tipler, and John A. Wheeler. 1988. *The Anthropic Cosmological Principle.* New York: Oxford University Press.

Bartlett, Frederic C. 1995. *Remembering: A Study in Experimental and Social Psychology.* Cambridge: Cambridge University Press.

Beit-Hallahmi, Benjamin. 2007. Atheists: A Psychological Profile. In *The Cambridge Companion to Atheism*, edited by M. Martin. Cambridge: Cambridge University Press.

Bering, Jesse M. 2002. Intuitive Conceptions of Dead Agents' Minds: The Natural Foundations of Afterlife Beliefs as Phenomenological Boundary. *Journal of Cognition and Culture* 2:263–308.

———. 2006. The Folk Psychology of Souls. *Behavioral and Brain Sciences* 29:453–462.

Bering, Jesse M., and David F. Bjorklund. 2004. The Natural Emergence of Reasoning about the Afterlife as a Developmental Regularity. *Developmental Psychology* 40:217–233.

Bering, Jesse M., C. Hernández-Blasi, and David F. Bjorklund. 2005. The Development of "Afterlife" Beliefs in Secularly and Religiously Schooled Children. *British Journal of Developmental Psychology* 23:587–607.

Bering, Jesse M., and Becky D. Parker. 2006. Children's Attributions of Intentions to an Invisible Agent. *Developmental Psychology* 42:253–262.

Bloom, Paul. 2004. *Descartes' Baby: How Child Development Explains What Makes Us Human*. London: Heinemann.

Boyatzis, Chris J. 2005. Religious and Spiritual Development in Childhood. In *Handbook of the Psychology of Religion and Spirituality*, edited by R. F. Paloutzian and C. L. Park. New York: Guilford Press.

Boyer, Pascal. 2001. *Religion Explained: The Evolutionary Origins of Religious Thought*. New York: Basic Books.

Burdett, Emily Reed, and Justin L. Barrett. in preparation. Children's Intuitions of Memory in Divine, Human, and Non-Human Minds.

Byrne, Donn E. 1971. *The Attraction Paradigm*. New York: Academic Press.

Casler, Krista, and Deborah Kelemen. 2008. Developmental Continuity in Teleo-Functional Explanation: Reasoning about Nature among Romanian Romani Adults. *Journal of Cognition and Development* 9:340–362.

Chandler, Michael J., and David Helm. 1984. Developmental Changes in the Contribution of Shared Experience to Social Role-Taking Competence. *International Journal of Behavioral Development* 7:145–156.

Chesterton, Gilbert K. 1955. *The Everlasting Man*. New York: Image Books.

Chilcott, Travis, and Raymond F. Paloutzian. 2010. The Cultivation of Divine Intimacy and Its Relation to Anthropomorphic Attribution: An Experimental Ethnographic Study on the Cognitive Effects of Gaudiya Vaishnava Religious Practices and Beliefs. Paper presented at the Cognition, Religion, and Theology Conference. University of Oxford, June 28–July 1, 2010.

Cohen, Emma. 2007. *The Mind Possessed: The Cognition of Spirit Possession in an Afro-Brazilian Religious Tradition.* New York: Oxford University Press.

Conway Morris, Simon. 2003. *Life's Solution: Inevitable Humans in a Lonely Universe.* Cambridge: Cambridge University Press.

Corkum, Valerie, and Chris Moore. 1998. Origins of Joint Visual Attention in Infants. *Developmental Psychology* 34:28–38.

Cosmides, Leda, and John Tooby. 1989. Evolutionary Psychology and the Generation of Culture, Part 2: Case Study: A Computational Theory of Social Exchange. *Ethology and Sociobiology* 10:51–97.

Csibra, Gergely. 2008. Goal Attribution to Inanimate Agents by 6.5-Month-Old Infants. *Cognition* 107:705–717.

Dawkins, Richard. 1986. *The Blind Watchmaker: Why the Evidence of Evolution Reveals a Universe without Design.* New York: Norton.

———. 2006. *The God Delusion.* London: Bantam Press.

Dennett, Daniel C. 2006. *Breaking the Spell: Religion as a Natural Phenomenon.* London: Allen Lane.

Ellsworth, Christine P., Darwin W. Muir, and Sylvia M. J. Hains. 1993. Social Competence and Person-Object Differentiation: An Analysis of the Still-Face Effect. *Developmental Psychology* 29:63–73.

Emmons, Robert A. 1999. *The Psychology of Ultimate Concerns: Motivation and Spirituality in Personality.* London: Guildford Press.

———. 2008. *Thanks! How Practicing Gratitude Can Make You Happier.* New York: Houghton Mifflin.

Emmons, Robert A., and Michael E. McCullough, eds. 2004. *The Psychology of Gratitude.* New York: Oxford University Press.

Evans, E. Margaret. 2001. Cognitive and Contextual Factors in the Emergence of Diverse Belief Systems: Creation versus Evolution. *Cognitive Psychology* 42:217–266.

Evans-Pritchard, Edward E. 1976. *Witchcraft, Oracles and Magic Among the Azande.* Oxford: Clarendon Press.

Flavell, John H., Eleanor R. Flavell, Frances L. Green, and Louis J. Moses. 1990. Young Children's Understanding of Fact Beliefs versus Value Beliefs. *Child Development* 61:915–928.

Flavell, John H., Patricia H. Miller, and Scott A. Miller. 1993. *Cognitive Development.* Englewood Cliffs, NJ: Prentice Hall.

Flew, Antony. 2007. *There Is a God: How the World's Most Notorious Atheist Changed His Mind.* New York: HarperOne.

Freud, Sigmund. 1983. *Totem and Taboo.* London: Ark Paperback.

———. 1989. *The Future of an Illusion.* Translated by J. B. Strachey. London: Norton.

Friedman, Batya. 1995. It's the Computer's Fault: Reasoning about Computers as Human Agents. In *Proceedings of the 2004 Conference on Human Factors in Computing Systems,* ACM Press (pp. 226–227).

Gelman, Susan A. 1988. The Development of Induction within Natural Kind and Artifact Categories. *Cognitive Psychology* 20:87–90.

Gelman, Susan A., and Kathleen E. Kremer. 1991. Understanding Natural Cause: Children's Explanations of How Objects and Their Properties Originate. *Child Development* 62:396–414.

Gergely, György, and Gergely Csibra. 2003. Teleological Reasoning in Infancy: The Naive Theory of Rational Action. *Trends in Cognitive Sciences* 7:287–292.

Gergely, György, Zoltán Nádasdy, Gergely Csibra, and Szilvia Bíró. 1995. Taking the Intentional Stance at 12 Months of Age. *Cognition* 56:165–193.

Giménez-Dasí, Marta, Silvia Guerrero, and Paul L. Harris. 2005. Intimations of Immortality and Omniscience in Early Childhood. *European Journal of Developmental Psychology* 2:285–297.

Goldman, Ronald G. 1964. *Religious Thinking from Childhood to Adolescence.* London: Routledge and Kegan Paul.

Granqvist, Pehr. 2005. Building a Bridge between Attachment and Religious Coping: Tests of Moderators and Mediators. *Mental Health, Religion, and Culture* 8:35–47.

Granqvist, Pehr, and Lee A. Kirkpatrick. 2004. Religious Conversion and Perceived Childhood Attachment: A Meta-Analysis. *International Journal for the Psychology of Religion* 14:223–250.

Gregersen, Peter K., Elena Kowalsky, Nina Kohn, and Elizabeth West Marvin. 2001. Early Childhood Music Education and Predisposition to Absolute Pitch: Teasing Apart Genes and Environment. *American Journal of Medical Genetics* 98:280–282.

Guthrie, Stewart E. 1993. *Faces in the Clouds: A New Theory of Religion.* New York: Oxford University Press.

Haidt, Jonathan. 2007. The New Synthesis in Moral Psychology. *Science* 316:998–1002.

Harris, Paul L. 2000. *The Work of the Imagination.* Oxford: Blackwell.

Harris, Paul L., Emma Brown, Crispin Marriott, Semantha Whittall, and

Sarah Harmer. 1991. Monsters, Ghosts and Witches: Testing the Limits of the Fantasy-Reality Distinction in Youth Children. *British Journal of Developmental Psychology* 9:105–123.

Harris, Paul L., and Marta Giménez. 2005. Children's Acceptance of Conflicting Testimony: The Case of Death. *Journal of Cognition and Culture* 5:143–164.

Harris, Paul L., and Melissa A. Koenig. 2006. Trust in Testimony: How Children Learn about Science and Religion. *Child Development* 77:505–524.

Hauser, Marc D. 2006. *Moral Minds: How Nature Designed Our Universal Sense of Right and Wrong.* New York: Ecco/HarperCollins.

Heider, Fritz, and Marianne Simmel. 1944. An Experimental Study of Apparent Behavior. *American Journal of Psychology* 57:243–249.

Henrich, Joseph, and Robert Boyd. 1998. The Evolution of Conformist Transmission and Between-Group Differences. *Evolution and Human Behavior* 19:215–241.

Henrich, Joseph, and Francisco Gil-White. 2001. The Evolution of Prestige: Freely Conferred Status as a Mechanism for Enhancing the Benefits of Cultural Transmission. *Evolution and Human Behavior* 22:165–196.

Hinde, Robert A. 1999. *Why Gods Persist: A Scientific Approach to Religion.* London: Routledge.

Hitchens, Christopher. 2007. *God Is Not Great: How Religion Poisons Everything.* New York: Twelve.

Humphrey, Nicholas. 2002. *The Mind Made Flesh: Frontiers of Psychology and Evolution.* New York: Oxford University Press.

Inagaki, Kayoko, and Giyoo Hatano. 2002. *Young Children's Naive Thinking about the Biological World.* New York: Psychology Press.

Jeeves, Malcolm, and Warren S. Brown. 2009. *Neuroscience, Psychology, and Religion: Illusions, Delusions, and Realities about Human Nature.* West Conshohocken, PA: Templeton Press.

Johnson, Marcia K., John D. Bransford, and Susan K. Solomon. 1973. Memory for Tacit Implications of Sentences. *Journal of Experimental Psychology* 98:203–205.

Johnson, Susan, Virginia Slaughter, and Susan Carey. 1998. Whose Gaze Will Infants Follow? The Elicitation of Gaze-Following in 12-Month-Olds. *Developmental Science* 1:233–238.

Katz, Leonard D., ed. 2000. *Evolutionary Origins of Morality: Cross-Disciplinary Perspectives.* Thoverton, UK: Imprint Academic.

Keil, Frank C. 1989. *Concepts, Kinds, and Cognitive Development.* Cambridge, MA: MIT Press.

Kelemen, Deborah. 1999. The Scope of Teleological Thinking in Preschool Children. *Cognition* 70:241–272.

———. 1999. Why Are Rocks Pointy? Children's Preference for Teleological Explanations of the Natural World. *Developmental Psychology* 35:1440–1453.

Kelemen, Deborah, and Cara DiYanni. 2005. Intuitions about Origins: Purpose and Intelligent Design in Children's Reasoning about Nature. *Journal of Cognition and Development* 6:3–31.

Kelemen, Deborah, and Evelyn Rosset. 2009. The Human Function Compunction: Teleological Explanation in Adults. *Cognition* 111:138–143.

Kendall-Tackett, Kathleen A., Linda Meyer Williams, and David Finkelhor. 2001. Impact of Sexual Abuse on Children: A Review and Synthesis of Recent Empirical Studies. In *Children and the Law: The Essential Readings*, edited by R. Bull. Oxford: Blackwell.

Kirkpatrick, Lee A. 2005. *Attachment, Evolution, and the Psychology of Religion.* New York: Guildford Press.

Knickmeyer, Rebecca C., and Simon Baron-Cohen. 2006. Foetal Testosterone and Sex Differences in Typical Social Development and in Autism. *Journal of Child Neurology* 21:825–845.

Knight, Nicola. 2008. Yukatek Maya Children's Attributions of Beliefs to Natural and Non-Natural Entities. *Journal of Cognition and Culture* 8:235–243.

Knight, Nicola, Paulo Sousa, Justin L. Barrett, and Scott Atran. 2004. Children's Attributions of Beliefs to Humans and God: Cross-Cultural Evidence. *Cognitive Science* 28:117–126.

Lanman, Jonathan A. 2009. *A Secular Mind: Towards a Cognitive Anthropology of Atheism.* Unpublished doctoral thesis, University of Oxford.

Lewis, C. S. 1960. *Mere Christianity.* New York: Macmillan.

Makris, Nikos, and Dimitris Pnevmatikos. 2007. Children's Understanding of Human and Supernatural Minds. *Cognitive Development* 22:365–375.

McCauley, Robert N. 2011. *Why Religion Is Natural and Science Is Not.* New York: Oxford University Press.

McCullough, Donald. 2005. *If Grace Is So Amazing, Why Don't We Like It?* San Francisco: Jossey-Bass.

McGrath, Alister. 2005. *Dawkins' God: Genes, Memes, and the Meaning of Life.* Oxford: Blackwell.

Meltzoff, Andrew N., and N. Keith Moore. 1983. Newborn Infants Imitate Adult Facial Gestures. *Child Development* 54:702–709.

Molina, Michèle, Gretchen A. Van de Walle, Kirsten Condry, and Elizabeth S. Spelke. 2004. The Animate-Inanimate Distinction in Infancy: Developing Sensitivity to Constraints on Human Actions. *Journal of Cognition and Development* 5:399–426.

Moon, Youngme, and Clifford Nass. 1998. Are Computers Scapegoats? Attributions of Responsibility in Human-Computer Interaction. *International Journal of Human-Computer Studies* 49:78–94.

Myers, David G. 2008. *A Friendly Letter to Skeptics and Atheists: Musings on Why God Is Good and Faith Isn't Evil.* San Francisco: Jossey-Bass.

———. 2009. *Psychology.* New York: Freeman.

Newman, George E., Frank C. Keil, Valerie Kuhlmeier, and Karen Wynn. 2010. Early Understanding of the Link between Agents and Order. *Proceedings of the National Academy of Sciences of the United States of America* 107:17140–17145.

Paloutzian, Raymond F. 1996. *Invitation to the Psychology of Religion.* Boston: Allyn & Bacon.

Pargament, Kenneth I. 1997. *The Psychology of Religion and Coping: Theory, Research, Practice.* London: Guilford Press.

Petrovich, Olivera. 1997. Understanding of Non-Natural Causality in Children and Adults: A Case against Artificialism. *Psyche en Geloof* 8:151–165.

———. 1999. Preschool Children's Understanding of the Dichotomy between the Natural and the Artificial. *Psychological Reports* 84:3–27.

Piaget, Jean. 1960. *The Child's Conception of the World*, translated by A. Tomlinson. Paterson, NJ: Littlefield, Adams.

Pinker, Steven. 1995. *The Language Instinct: How the Mind Creates Language.* New York: HarperPerennial.

———. 2002. *The Blank Slate: The Modern Denial of Human Nature.* New York: Viking.

Reid, Thomas. 1983. *Inquiry and Essays*, edited by R. E. Beanblossom and K. Lehrer. Indianapolis, IN: Hackett.

———. 1997. *Inquiry into the Human Mind on the Principles of Common Sense*, edited by D. R. Brookes. Edinburgh: Edinburgh University Press.

Reisberg, Daniel. 1997. *Cognition: Exploring the Science of the Mind.* New York: Norton.

Richert, Rebekah A., and Justin L. Barrett. 2005. Do You See What I See?

Young Children's Assumptions about God's Perceptual Abilities. *International Journal for the Psychology of Religion* 15:283–295.

Rochat, Philippe, Rachel Morgan, and Malinda Carpenter. 1997. Young Infants' Sensitivity to Movement Information Specifying Social Causality. *Cognitive Development* 12:537–561.

Rochat, Philippe, Tricia Striano, and Rachel Morgan. 2004. Who Is Doing What to Whom? Young Infants' Developing Sense of Social Causality in Animated Displays. *Perception* 33:355–369.

Saler, Benson, and Charles A. Ziegler. 2006. Atheism and Apotheosis of Agency. *Temenos* 42:7–41.

Scholl, Brian, and Patrice D. Tremoulet. 2000. Perceptual Causality and Animacy. *Trends in Cognitive Sciences* 4:299–308.

Simons, Daniel J., and Frank C. Keil. 1995. An Abstract to Concrete Shift in the Development of Biological Thought: The Insides Story. *Cognition* 56:129–163.

Slone, D. Jason. 2004. *Theological Incorrectness: Why Religious People Believe What They Shouldn't.* New York: Oxford University Press.

Spelke, Elizabeth S., and Katherine D. Kinzler. 2007. Core Knowledge. *Developmental Science* 11:89–96.

Spelke, Elizabeth S., Ann Phillips, and Amanda L. Woodward. 1995. Infant's Knowledge of Object Motion and Human Action. In *Causal Cognition: A Multidisciplinary Debate*, edited by D. Sperber, D. Premack, and A. J. Premack. Oxford: Oxford University Press.

Sperber, Dan. 1996. *Explaining Culture: A Naturalistic Approach.* Oxford: Blackwell.

Sperber, Dan, David Premack, and Ann James Premack, eds. 1995. *Causal Cognition: A Multidisciplinary Debate.* New York: Oxford University Press.

Spykman, Gordon. 1992. *Reformational Theology: A New Paradigm for Doing Dogmatics.* Grand Rapids, MI: Eerdmans.

Swinburne, Richard. 1996. *Is There a God?* Oxford: Oxford University Press.

———. 2008. *Was Jesus God?* Oxford: Oxford University Press.

Taylor, Marjorie. 1999. *Imaginary Companions and the Children Who Create Them.* New York: Oxford University Press.

Tomasello, Michael. 1999. *The Cultural Origins of Human Cognition.* Cambridge, MA: Harvard University Press.

Tremlin, Todd. 2006. *Minds and Gods: The Cognitive Foundations of Religion.* New York: Oxford University Press.

Trigg, Roger. 2007. *Religion in Public Life: Must Faith Be Privatized?* Oxford: Oxford University Press.

Tversky, Amos, and Daniel Kahneman. 2000. Judgments under Uncertainty: Heuristics and Biases. In *Judgment and Decision Making: An Interdisciplinary Reader*, edited by T. Connoly, H. R. Arkes, and K. R. Hammond. Cambridge: Cambridge University Press.

Wellman, Henry, David Cross, and Julanne Watson. 2001. Meta-Analysis of Theory of Mind Development: The Truth about False-Belief. *Child Development* 72:655–684.

Wellman, Henry M., and David Estes. 1986. Early Understanding of Mental Entities: A Reexamination of Childhood Realism. *Child Development* 57: 910–923.

Wellman, Henry M., and Susan A. Gelman. 1998. Knowledge Acquisition in Foundational Domains. In *Handbook of Child Psychology*, edited by W. Damon. Hoboken, NJ: Wiley.

Whitehouse, Harvey. 1995. *Inside the Cult: Religious Innovation and Transmission in Papua New Guinea.* Oxford: Clarendon Press.

Whitehouse, Harvey. 1996. Apparitions, Orations, and Rings: Experience of Spirits in Dadul. In *Spirits in Culture, History, and Mind*, edited by J. Mageo and A. Howard. New York: Routledge.

Wigger, J. Bradley. 2010. Imaginary Companions, Theory of Mind, and God. Paper presented at the Cognition, Religion, and Theology Conference, University of Oxford, June 29, 2010.

———. 2011. See-Through Knowing: Learning from Children and Their Invisible Friends. *Journal of Childhood and Religion* 2.

Winograd, Eugene, and Ulric Neisser, eds. 1992. *Affect and Accuracy in Recall: Studies of "Flashbulb" Memories.* Cambridge: Cambridge University Press.

Xinzhong, Yao, and Paul Badham. 2007. *Religious Experience in Contemporary China.* Cardiff: University of Wales Press.

Zaitchik, Deborah. 1990. When Representations Conflict with Reality: The Preschooler's Problem with False Beliefs and "False" Photographs. *Cognition* 35:41–68.

Zia Ullah, Mohammad. 1984. *Islamic Concept of God.* London: Kegan Paul.

INDEX

About the Author

Justin L. Barrett earned degrees in psychology from Calvin College (B.A.) and Cornell University (Ph.D.). He served on the psychology faculties of Calvin College and the University of Michigan (Ann Arbor) and as a research fellow of the Institute for Social Research. Barrett is an editor of the *Journal of Cognition and Culture* and is the author of numerous articles and chapters concerning the cognitive science of religion. His book *Why Would Anyone Believe in God*? (2004) presents a scientific account for the prevalence of religious beliefs. He is research associate at Oxford's Centre for Anthropology and Mind in the School of Anthropology and Museum Ethnography, and holds the Thrive Chair of Applied Developmental Science at Fuller Theological Seminary's School of Psychology. Barrett lives in Pasadena, California, with his wife, Sherry Barrett, and his son, Skylar, and daughter, Sierra.